GEORGE STRACHAN OF THE MEARNS

Scottish Religious Cultures *Historical Perspectives*

Series Editors: Scott R. Spurlock and Crawford Gribben

Religion has played a key formational role in the development of Scottish society shaping cultural norms, defining individual and corporate identities, and underpinning legal and political institutions. This series presents the very best scholarship on the role of religion as a formative and yet divisive force in Scottish society and highlights its positive and negative functions in the development of the nation's culture. The impact of the Scots diaspora on the wider world means that the subject has major significance far outwith Scotland.

Available titles

George Mackay Brown and the Scottish Catholic Imagination
Linden Bicket

Poor Relief and the Church in Scotland, 1560–1650
John McCallum

Jewish Orthodoxy in Scotland: Rabbi Dr Salis Daiches and Religious Leadership
Hannah Holtschneider

Miracles of Healing: Psychotherapy and Religion in Twentieth-century Scotland
Gavin Miller

George Strachan of the Mearns: Seventeenth-century Orientalist
Tom McInally

Forthcoming titles

The Scot Afrikaners: Identity Politics and Intertwined Religious Cultures
Retief Muller

Dugald Semple and the Life Reform Movement
Steven Sutcliffe

Presbyterianism Re-established: The Presbyteries of Dunblane and Stirling after the Williamite Revolution
Andrew Muirhead

William Guild and Moderate Divinity in Early Modern Scotland
Russell Newton

The Dynamics of Dissent: Politics, Religion and the Law in Restoration Scotland
Neil McIntyre

The Catholic Church in Scotland: Financial Development 1772–1930
Darren Tierney

edinburghuniversitypress.com/series/src

GEORGE STRACHAN OF THE MEARNS

Seventeenth-century Orientalist

TOM McINALLY

EDINBURGH
University Press

Edinburgh University Press is one of the leading university presses in the UK. We publish academic books and journals in our selected subject areas across the humanities and social sciences, combining cutting-edge scholarship with high editorial and production values to produce academic works of lasting importance. For more information visit our website: edinburghuniversitypress.com

First published in hardback by Edinburgh University Press 2020

Edinburgh University Press Ltd
The Tun – Holyrood Road
12 (2f) Jackson's Entry
Edinburgh EH8 8PJ

Typeset in 10/12 ITC New Baskerville by
Servis Filmsetting Ltd, Stockport, Cheshire

A CIP record for this book is available from the British Library

ISBN 978 1 4744 6622 6 (hardback)
ISBN 978 1 4744 6623 3 (paperback)
ISBN 978 1 4744 6624 0 (webready PDF)
ISBN 978 1 4744 6625 7 (epub)

Contents

Acknowledgements vi

Introduction 1

1 Heritage 8

2 Exile 19

3 The Humanist Scholar 29

4 To Constantinople 40

5 Aleppo 53

6 Mohammed Çelebi 66

7 The Ḥusaynābādī Scholiasts 78

8 Strachan's Library 88

9 The English East India Company 98

10 'Stracan our Infernall Phesition' 113

11 Among Friends 129

12 The Mission at Srinagar 143

Appendix 157

Archives 183

Bibliography 184

Index 193

Acknowledgements

My interest in George Strachan derived from my doctoral dissertation on the Scots Colleges abroad. The brief entry in the register of the Pontifical Scots College in Rome sent me on a programme of research into what, to me, is his fascinating life story. Much of my initial research drew on the material gained in the Vatican Archives (ASV), Vatican Library (BAV) and the Jesuit Archives in Rome (ARSI). This soon led to researching Strachan's album amicorum and Della Valle's journals, for which I drew on the services of my good friends in the Special Collections of Aberdeen University Library. I am in debt to all of the staff of these archives for their help. Also, I thank my colleagues in the Research Institute for Irish and Scottish Studies in the University of Aberdeen for the support and encouragement they gave me, as I researched areas beyond my normal field of study, and my friend Colin Chapman for his help in correcting the final text. Lastly my family, particularly my ever tolerant wife, deserves a special thank you for putting up with my obsession with George Strachan over several years.

Introduction

Call for a Biography

In 1983 Victor Winstone, the noted writer on the Middle East and member of the Royal Geographic Society, delivered a paper entitled 'George Strachan, 17th Century Orientalist' at a seminar for Arabian Studies held in London. He subtitled his paper 'Plea for a Biographical Study' (Winstone 1984: 103–9). A noted biographer himself, Winstone described Strachan as one of the greatest oriental scholars of his time. Although there had been some investigation into the life of Strachan, notably that by Giorgio Levi Dellavida, professor of Semitic languages in Rome (Dellavida 1956), and an early paper by Fr David McRoberts (McRoberts 1952: 110–28) it surprised Winstone that historians had still to conduct a full study of such a deserving subject. He felt that the omission was in large part due to the fact that Strachan had left no record of his travels and, as an earlier researcher had written, 'his footsteps [could be] tracked piecemeal, only as the palaeontologist makes out the intermittent traces of an extinct wader or batrachians upon the petrified mud of the Eocene' (Yule 1888: 312).

Research into oriental studies has grown greatly since the work of Johann Fück (Fück 1955) and in the twenty-first century numerous scholars have added substantially to the body of work through such noted series as *The History of Oriental Studies,* published in Leiden, but there has still been no new biography of Strachan. The late Professor Bosworth included a chapter, ten pages long, summarising the most prominent facts known of Strachan's life and work (Bosworth 2012). In the period since Winstone made his plea, however, some further evidence of Strachan's life has come to light, but most of what is known derives from accounts given by others who crossed the Scotsman's path. Even with this additional knowledge it is still impossible to provide the full biographical account that Winstone felt was justified. Nevertheless, a more rounded picture is emerging of this remarkable man.

George Strachan was a humanist scholar and a member of the European-wide Republic of Letters during a period when academic institutions were experiencing significant growth and intellectual dispute was intense on both religious and philosophical grounds. He made a reputation for himself writing Latin poetry of considerable quality, but his greatest contribution to early modern scholarship was as an orientalist whose knowledge of Middle Eastern languages greatly exceeded that of any other European scholar in the first half of the seventeenth century. He spoke at least eight European and Eastern languages fluently but it can also be claimed that he was the first Western scholar to achieve a true understanding of Arabic and Persian texts free from European misinterpretations.

European Interest in the East

European Christian scholars had possessed an understanding of the languages and cultures of the East from the times of the Crusades and earlier. Pope Sylvester II (c. 946–1003) promoted studies of Graeco-Roman sciences through Arabic sources. He was inspired in his own scientific work by the Islamic educational institution of Cordoba. Western knowledge of the Muslim world progressed through access to the libraries of Arab Spain. In the thirteenth century King Alfonso X ('El Sabio') of Castile (r. 1252–84) sponsored scholars to translate Arabic texts into Castilian (Doubleday 2015). In a number of cases these texts had initially been translated into Hebrew by Jewish scholars. Scientific texts were further translated into Latin and circulated abroad. Christoph Clavius (1538–1612), a German Jesuit, used Latin translations of Arabic accounts of Aristotle and Arab mathematicians and astronomers in his writings citing, among others, Ibn Rushd, Thābit b. Qurra, Abū Ma'shar, al-Biṭrūjī and al-Farghānī as sources in his published works (Knobloch 2002: 257–84).

Additional knowledge was gained through the extensive trading that took place between Italian city states, principally Venice and Genoa, and the developing Ottoman Empire and Mamluk Egypt. With the fall of Constantinople in 1453 Western scholars gained access to classical manuscripts brought to the West by Greek scholars. These were in a variety of languages and in addition to works by ancient writers their libraries contained scholarly texts which shed a better light on the work of Arab scholars. Given the circuitous routes that such books had taken to reach the West over the centuries, it was inevitable that unwitting simplifications and errors had crept into the translations available to Christian scholars. Even if the texts had been flawless, their mere possession would not have ensured that they were correctly interpreted: an understanding of the context of the society and culture in which the language was used was also necessary. For the most part, prior to the sixteenth century, Western scholars lacked such knowledge. In addition they rarely had the opportunity to converse with native speakers of the languages of these texts.

The lack of proficiency in Eastern languages was of concern to more than the academic community. Commercial ventures of the Western trading states were burgeoning with the Ottoman Empire and other Eastern states. It was unusual, however, for merchants to have more than the limited proficiency in the languages of their trading partners needed to conduct business. Merchants who traded in the Eastern Mediterranean and Ottoman lands circumvented the problem by developing a lingua franca based largely on Italian, which came to be used by Muslim as well as Christian traders (see Chapter 5). However, Christian traders were left at a disadvantage in their dealings with the state authorities, where it was essential to have a deeper understanding not only of the languages but of the protocol to be observed at oriental courts. The Venetians recognised the danger their deficiency in this regard represented for their mercantile interests when they were engaged in negotiating contracts with the Ottoman sultans. They were concerned that the dragomans they were forced to use as interpreters and intermediaries were not willing to present arguments as forcefully as a Venetian would. In his report to the Venetian *signoria* (governing body) in 1576, the *bailo* (Venetian ambassador) in Constantinople, Antonio Tiepolo, complained that the dragomans, who were Turkish subjects, were afraid to express the Venetian position robustly in their negotiations with the court officials (Albèri 1839: 185). In addition, the *bailo* suspected that the dragomans were prepared to interpret terms of negotiation to their own financial advantage (Gürkan 2015: 110–12). In light of these concerns, in the late sixteenth century the *signoria* agreed to shoulder the expense of establishing a Turkish and Arabic language school in their trading depot in Constantinople. The facility was set up to train interpreters whom they could trust to conduct diplomatic and commercial negotiations (Lucchetta 1989: 19–40). This commercial enterprise can be considered as a serious effort by a Western power to establish a permanent facility providing practical education in Eastern languages using native speakers. The attempt, however, failed. Linguistic proficiency was only part of the training needed and it took years of familiarisation to understand the customs and politics of the Ottoman court. The *bailo*'s term of office was limited to two years and very few members of the Venetian nobility were willing to spend the time needed to acquire an in-depth knowledge of Ottoman society, knowing that it would require them to be away from Venice for the greater part of their careers. The school's failure was due primarily to a lack of qualified interested candidates (Lucchetta 1989: 25–6).

Other Christian nations were slow to realise the disadvantages of neglecting their Eastern neighbours' languages. Early interpretations of Arabic texts by the scholars of Western universities, all of which were Church institutions, were hindered by their refusal to view favourably any writings inspired by the 'false religion' of Islam. In Spain the Inquisition confiscated Arabic books in its efforts to eradicate signs of the country's Islamic past. Western study of the Qur'an was carried out with the purpose of discrediting its contents and not as a means of learning Arabic. The first Latin translation made by Robert of Ketton in 1142–3 was full of inaccuracies and remained the only one available for over three centuries. It led generations of Western scholars to have at best an incomplete and at worst an erroneous understanding of this text (Burman 1998: 703–32). Among the false ideas Western religious scholars held was that Islam was a religion of lust, citing polygamy and the seclusion of women as proof (Norman 1960: 118–25).

The first Western attempt to understand the Qur'anic text rather than denigrate it was commissioned by Cardinal Egidio da Viterbo in 1518. It was undertaken by two converts from Islam, Juan Gabriel and Leo Africanus, but was never published. By circulating in manuscript form it was available only to a limited number of readers and was not used for teaching the Arabic language. There is no evidence that the cardinal read the translation he had commissioned. Although an acknowledged scholar, his interest was in the Cabbala and his proficiency in Eastern languages did not extend to learning Arabic (Martin 1992: 157).

Later, through the work of Protestant reformers, attitudes to learning Arabic began to change but at the time George Strachan was a student in Europe the study of Islamic Arabic texts was severely circumscribed (Hamilton 2001: 169–82). These failures prevented Christian scholars from participating in a thorough academic discussion of the religion, philosophy or culture of the Middle East.

This refusal to engage fully with the work of Eastern scholars became increasingly dangerous throughout the sixteenth century due to the growing power of the Muslim states. When Sultan Selim I conquered Mamluk Egypt and incorporated it into his empire in 1517, the Ottomans became the principal Muslim power in the Middle East. The sultan became the caliph of all Sunni Muslims and protector of the holy sites of Mecca, Medina and Jerusalem. Even before this considerable expansion of their power, the Ottoman sultans posed a significant military threat to Christian Europe. Later, under Suleiman the Magnificent, the empire was extended further by territorial gains in Europe and the Christian kingdoms of the Caucasus. Successive popes encouraged European countries most at risk to form defensive leagues and exhorted all Christians (Catholic, Orthodox and Protestant) to provide aid to fellow Christians under attack by Muslim armies.

Due to the importance of its trade with the East, which earned it the soubriquet of 'the Turk's Courtesan', the Venetian Republic normally remained outside such leagues and therefore was often treated with disdain by the main body of Christendom (Dursteler 2006: 5). However, in the mid-sixteenth century it was forced to join with other Western powers against the Ottomans in a series of wars which met with mixed success. When a peace treaty was agreed in 1573, Venice withdrew from all military leagues and adopted a position of armed neutrality with the Ottoman Empire (Valerio 1679: 2). A number of other nations, influenced by the degree to which their territories were at threat from Ottoman conquest, withheld support from the leagues and followed the diplomatic route taken by Venice (Agoston 2007: 75–103). This inaugurated a lengthy period of peace between the Most Serene Republic (La Serenissima) and the Sublime Porte (Âsitâne-yi Sa'âdet) which lasted until 1645 and allowed commercial and diplomatic contacts between them to flourish.

During this period of peace, the Roman Catholic Church recognised its need for greater understanding of Eastern languages and cultures. The Church's missionary activity had developed greatly in the East and it became essential that recruits to the missions be given relevant training in the languages and cultures of the ever-increasing number of new countries in which the Church worked. In order to meet this requirement, in 1610 Pope Paul V decreed that a number of monasteries and convents in Italy should become centres of instruction and research into Eastern languages. Despite the mandatory nature of the pope's decree, only the Order of the Caracciolini embraced it with any enthusiasm, having been running such a school in Rome since 1595. The papal remit to the new schools was that they should gain mastery of Turkish and Arabic, in recognition of their being the languages of the most dangerous enemy of Christendom, but also of Persian and the languages of the Indian subcontinent, Southeast Asia, Japan and China. The missionaries were to be taught the languages they needed, together with the ability to refute Islamic texts in order to gain converts (Zwartjes 2012: 185–242).

The new missionaries were needed to follow the expansion of the trading empires established by the Portuguese and Spanish (Bernardini 2011: 265–81). Following the initial exploratory voyages of Vasco da Gama at the end of the fifteenth century, the Portuguese had traded east of the Cape of Good Hope for nearly one hundred years without serious European competition. Spain's arrival in the Far East in the mid-sixteenth century to trade with Japan and the Philippines did not create a rivalry, at least in theory. When the two kingdoms were united under King Philip II/I of Spain and Portugal in 1580, a degree of cooperation between the two colonial powers was expected. The papacy saw the potential for the establishment of missions throughout the region which, in addition to the advances being made in the New World, allowed Rome to view its work as that of a universal (Catholic) church.

Tracing George Strachan

It was against this background that the Scotsman George Strachan began to apply his talents to gain an understanding of Eastern languages. In order to learn from native speakers and gain access to literature unobtainable in the West, he undertook an extensive journey throughout much of the Middle East. As a youth, he had a wanderlust coupled with a fondness and natural ability for learning languages. The papal exhortation to European scholars to gain a better understanding of Eastern languages would have been all that was necessary to inspire the man from the Mearns to engage in serious research. In 1613 George Strachan set off on what can be viewed as a pilgrimage through the Ottoman, Persian and Mughal Empires. While on his peregrinations he collected important books in Arabic, Turkish and Persian which had been hitherto unknown, unavailable or poorly understood in the West. He studied them, often under the tutelage of eminent Islamic scholars, gaining an unequalled understanding which he endeavoured to share through his translations. He added extensive annotations and glosses to the texts which gave nuanced explanations of difficult passages. It was always his intention to make this knowledge available to Western scholars as he showed by sending much of his library westwards to Rome. By doing so, he provided European scholars with the means not only of developing a greater understanding of the subtleties of the languages but also of gaining a deeper insight into those Eastern societies through their cultures of science, philosophy and religion.

He achieved the monumental goal of collecting, translating and interpreting these books while engaged on extensive travels. He took advantage of opportunities afforded by Western travellers returning home to send both letters to his friends and his growing library to Rome. He never returned to Europe, although initially it was his intention to do so, but through his actions many of the volumes that he had accumulated were safely preserved in the libraries of the teaching convents and used for the benefit of both seasoned scholars and students. Some of his books have since been lost but many have survived in European archives. The range of subjects and Strachan's explanations of the texts demonstrate his pre-eminence among his contemporaries as a scholar of oriental languages. Also, when considering the efforts he made to acquire his library, the modern researcher cannot fail to be impressed by the heroic nature of Strachan's journeys in the East.

Despite the availability of this material, the Scotsman remains a shadowy figure. The historiography relating to Strachan is extremely limited. The only non-contemporary biography is the slim volume which Giorgio Dellavida wrote at the request of the Third Spalding Club. The work was commissioned in the 1930s but was published nearly twenty years later. The long gestation period was due in part to the disruption caused by the Second World War but also to the difficulty the author faced in unearthing relevant material on his subject. In his introduction Dellavida commented on the paucity of information available and the limits which this had placed on his research. The work is, however, extremely valuable. The author, as well as providing an outline of the then known aspects of Strachan's life, has produced the most comprehensive catalogue extant of Strachan's library including the locations of the books and an assessment of the importance of the works as Arabic and Persian texts. Unfortunately none of his Turkish texts which are known to have existed have survived.

As Dellavida discovered, no individual archive or contemporary account throws much light on Strachan. Any research requires examination of a wide range of sources. For much of his life, George Strachan appears to be nearly invisible. Many of his details are known only from his involvement in the lives of others. As a humanist he wrote a great deal, including a considerable quantity of poetry, but little of what survives is about Strachan.

Accounts of him written by his contemporaries, although numerous, are rarely long, and each of these commentators has been able to describe only a small part of his career. Any researcher is obliged to piece together the narrative of George Strachan's life from a large number of disparate sources with little direct input from the man himself. The account which emerges is one where knowledge of the history of the archives involved becomes as important as the biography of Strachan that can be deduced from them.

There are, in total, seven major sources of information on Strachan. One has already been mentioned: the collection of books which he sent to Europe. A second is the album amicorum (book of friends) (SCA, CB/57/12) which Strachan used as a young scholar. A third can be found in references contained in Jesuit correspondence between Scotland and Rome which is kept in the Jesuit archives in Rome (ARSI) and the Vatican archives (ASV). His own writings – poems, book dedications and a few surviving letters, which provide a tantalising picture of the man although regrettably one of limited scope – constitute a fourth important archive (Leask 1910: 338–46). Thomas Dempster of Muiresk, who described himself as a close friend of Strachan, included an account of his fellow Catholic and his humanist writings in *Historia Ecclesiastica Gentis Scotorum*, a set of biographies of prominent Scots, published posthumously in Bologna in 1627. Another source of information is provided by the Italian nobleman Pietro Della Valle, who in his published letters gives an account of the life of Strachan in the Middle East having befriended the Scotsman in Persia (Della Valle 1664). The last source archive of consequence is that of the English East India Company for which Strachan worked in Persia and possibly India (*Calendar of State Papers* 1857).

The varied nature of these sources also has a bearing on the reliability to be placed on the details of the story being told. Commentators, even when they are being honest, may have been influenced by feelings of friendship or enmity. This can be seen in the different descriptions of the character of the man given by Della Valle and some of the officials of the English East India Company. In reading their accounts it is clear that factual errors in aspects of Strachan's life have been made, since contradictions exist. Identification of the true version of events is often impossible and the temptation to stray from argument to speculation is a constant danger for any researcher. There is, however, a distinction to be made between speculation and reasoned interpretation of the known facts. The writer hopes he has managed to maintain this distinction in describing Strachan's life. Although the multiplicity of sources is undoubtedly frustrating it does have the advantage of providing a more textured description of the man. One conclusion can be arrived at with certainty: Strachan's achievements are such that they can only be explained as the work of someone with extraordinary ability and a remarkable personality. His many friends have testified to these qualities but it is appropriate to begin an account of the man and his life with the little that Strachan has said about himself.

CHAPTER ONE

Heritage

Of Royal Descent

In his printed poems Strachan gives his name as Georgius Strachanus Merniensis Scotus – George Strachan of the Mearns, Scot. This scant information is supplemented by the family coat of arms on the cover of his album amicorum and by the comments written inside by his friends, professors and fellow students. From these it can be shown that Strachan was born c. 1572, the youngest of three sons of Sir Alexander Strachan of Thornton (d. c. 1600) and Isobel Keith (c. 1543–August 1595). Sir Alexander was the 12th Strachan of Thornton. George's mother was the daughter of William Keith, 4th Earl Marischal, and through his lineage Strachan and his siblings were direct descendants of King James I of Scotland (Balfour 1904: 46–7). Throughout his life Strachan placed great importance on his social status and, no matter how impecunious were the straits in which he found himself, he always expected to be treated with the respect due to a gentleman of noble descent. All parts of his extended family were nobility and gentry, holding lands which stretched from Strathdon in the north-east of Scotland to Dundee in the east. Dunottar Castle, the seat of the Earl Marischal, George's grandfather, is less than fifteen miles from the Strachans' family home, Thornton Castle. The Thornton estate lies in the rich farmlands of the Howe of the Mearns between the small towns of Laurencekirk and Fettercairn (Balfour 1904: 122).

In the late sixteenth century Scottish nobility and gentry were divided by religious confession. In 1591, when George was still a young man, there were sixteen 'Papists and discontented Erles and Lordes' and only eight 'Protestants and [those] well affected to the course of England' of similar status. Those nobility and gentry of inferior rank to earls and lords showed an opposite balance, with records for 1592 stating: 'Protestants 28, Papists 13, neutral, suspect or doubtful 6, minors 9' (Rogers 1873: 62–3). Strachan of Thornton was strongly Catholic. George's maternal grandfather, the Earl Marischal, was one of twelve peers chosen by Queen Mary in 1560, while still queen of France, to act in her absence as a governing council for Scotland following the death of her mother, Mary of Guise. George's eldest brother, Robert, was his father's heir and stood to inherit all of the family lands. Robert married Sarah Douglas, daughter of the Earl of Angus, in 1586 and shortly afterwards they had a son, Alexander (Balfour 1904: 122).

With the succession thus secured to the next generation, as a younger son George had little prospect of significant financial benefit from his family and would have been expected to make his own fortune.

Few career choices of appropriate social status were open to junior members of the minor nobility. Often younger sons took up positions at court, entered the officer corps of the army (when a standing army existed) or were ordained into the Church. None of these was a realistic option in Scotland for a Catholic in the late sixteenth century. Even obtaining a higher education was difficult for those who did not subscribe to the Calvinist Confession of Faith. Initially King's College, Aberdeen, was tolerant of non-conformists and allowed them to matriculate but it was impossible for them to graduate. However, George Strachan's name does not appear in the college records (Anderson 1893). Like most well-born Catholics in Scotland, George's early education would have been provided by a private tutor at home or in the house of some relative or family friend. From 1581 Jesuit missionaries were lodged with many of the Catholic nobility in Scotland and, as well as carrying out their priestly duties, they acted as tutors to the families of their hosts. In the north-east of Scotland they are known to have stayed in the Castles of Huntly, Strathbogie, Slaines and Letterfourie (O'Neill and Domínguez 2001: 1259–62). A report by Lord Burghley, William Cecil, Queen Elizabeth of England's Lord Privy Seal, in 1590 stated:

> all the Northern part of the Kingdom, including the shires of Inverness, Caithness, Sutherland, and Aberdeen, with Moray, and the Sherrifdoms of Buchan, of Angus, of Wigton, and of Nithsdale, were either wholly, or for the greater part, commanded mostly by noblemen who secretly adhered to that faith (Catholicism), and directed in their movements by Jesuits and Priests, who were concealed in various parts of the country, especially in Angus. (Gordon 1869: iii)

The Jesuits stationed in these houses would have been known to the Strachan family and have been willing to accept young George as a pupil. Following his initial schooling in Latin grammar, George's parents decided that he should go to France for his higher education. The exact date of his leaving is unknown but there are indications that it was about 1588, when George was aged sixteen, and about the same time as Robert's son, Alexander, was born. The arrival of the infant may have been the catalyst for the decision to educate the young man abroad. The family's choice of Paris as Strachan's place of study is the clearest indication of the date of 1588 being correct. Scots Catholics who wanted to take a course of higher studies had little option but to travel to mainland Europe and enrol at a college in a Catholic country. One of the most popular among Scottish students was the Northern College, one of a group of Catholic institutions of higher education at Braunsberg in Livonia (now Braniewo in north-east Poland) (Bender 1868: 15–16). It was established in 1578 by the authority of Pope Gregory XIII specifically for the education of Catholics from the Protestant countries of northern Europe. Students from Scandinavia as well as Scotland studied there. A group of Scottish Jesuits had helped set up and staff the university and in 1580 one of their number, Robert Abercrombie, recruited its first Scottish students and escorted them to Braunsberg. The college remained a centre of higher education for Scots until 1626, when the Swedish army of Gustavus IV Adolphus overran Braunsberg during the Thirty Years' War and closed the city's Catholic institutions. For more than four decades prior to this catastrophe, the Northern College educated Scots Catholics (Fischer 1902: 298–9).

Even with this facility available to them, Scots petitioned the pope for a college exclusively for themselves. With the political and financial patronage of Mary, Queen of Scots, they succeeded in setting up the queen's new college in Pont-à-Mousson in Lorraine with a Scottish Jesuit, William Crichton, as its rector. Following Mary's execution in 1587, the queen's uncles, the duke and the cardinal of Lorraine, continued to fund the college but in 1588, when they were assassinated, the college was forced to close and its students disperse to complete their studies elsewhere. Although it is reasonable to assume that George Strachan set out for France with the intention of studying at the Scots College in Pont-à-Mousson, there is no record of him in the college register (Anderson 1906). If he had left home earlier than 1588, he would have been able to enrol: if later, his family would have known of the closure of the college and would have advised him to go to the Northern College in Braunsberg. This indicates a likely date of 1588 or 1589 for Strachan's departure from Scotland.

Like other young Catholics seeking an education abroad, it would have been necessary for Strachan to keep his journey secret from the authorities of the State and Kirk. The Penal Laws forbade studying at Catholic colleges, and families who sent their sons abroad for this purpose were heavily fined if discovered (Hebermann 1913: 'Penal Laws'). Such journeys required careful planning and students travelling abroad normally did so in small groups led by a Jesuit missionary, an older family member or trusted friend. When possible their first destination in France was Paris which had long been a centre for Scots in Europe and had a substantial expatriate community. There they would have been able to call on the most senior member of the Scottish Catholic community in exile, James Beaton. Beaton was the last pre-Reformation archbishop of Glasgow and in 1560 had been forced to leave Scotland by the anti-Catholic laws passed that year by the Reformation Parliament. When Queen Mary left France for Scotland in 1561, she appointed Beaton as her ambassador to the court of her brother-in-law, Charles IX. Beaton held the position of Scottish ambassador to the French court until his death in 1603. Following his mother's execution, King James VI continued to use the archbishop in this capacity at the courts of the succeeding French kings, Henri III and Henri IV. The archbishop was well placed to offer information and advice on the affairs of

France to his visiting countrymen. He would also have been able to inform Strachan's party of the situation in Pont-à-Mousson, having been involved in the establishment of the college, and he was also in a position to recommend alternative facilities available for study in Paris. There was no Scots College in Paris at that time. The long desired college was established in 1603 and its creation became possible only on the death of Archbishop Beaton using the bequest he made of his house and estate. There were, however, other opportunities for Scots to study in the city.

Educated by Jesuits

During the second half of the sixteenth century, the Society of Jesus had gained the reputation of being an excellent provider of higher education. Jesuits were much sought after by the Church and civic authorities to set up and run academic colleges. The value placed on their skill as educators did not, however, overcome the distrust with which they were viewed in France. The source of this distrust was the allegiance that they held to the pope. Relations between the Gallican Church and the papacy were often strained. However, Scottish Jesuits were not subject to the same level of animosity as others in their Society, due in large part to the high regard in which the Queen of Scots was held in France. Expressions of sympathy which followed her execution were extended to all Scottish Catholics, even Jesuits. The long alliance between the two countries had ensured a favourable sentiment for Scots but the queen's imprisonment and the manner of her eventual execution strongly reinforced it. Mary was a queen of France as well as of Scotland. The Parisian public reacted to her death with revulsion and fury. The English ambassadors, Sir Edward Stafford and Sir William Wade, reported to Queen Elizabeth that when news of Mary's execution arrived in Paris neither of them dared venture onto the streets for fear of being attacked by the mob (Black 1959: 388).

Particularly strong Scottish Jesuit connections had developed with one of the numerous colleges in Paris. The Jesuit College (later known as Lycée Louis-le-Grand) had been endowed by the Bishop of Claremont, William du Prat, and was commonly known as Claremont College. On founding his college in 1563, Bishop William had appointed a Scottish Jesuit, Edmund Hay, as its first rector. Hay was succeeded by another Scot, John Tyrie, and the college continued to be run by Scots for much of the remainder of the century. Both Hay and Tyrie were colleagues of Robert Abercrombie and had been involved, along with him and William Crichton, the rector of the queen's college in Pont-à-Mousson, in starting up the colleges in Braunsberg. Throughout its early history, although it had not been established for them, Claremont College accepted Scots as students. When George Strachan and his party arrived in Paris and presented themselves to Beaton, it would have been natural for the archbishop to introduce them to the then principal of the college, Edmund Hay. Early student records for Claremont College have not survived, so it is impossible to substantiate Strachan's attendance, but there is strong evidence in the accounts of his education provided in his album amicorum that this was the case. Enrolment at Claremont would have meant that Strachan maintained his involvement with Jesuit tutors which is likely to have started with his early education at home in Scotland. He was to continue his association with the Society for the greater part of his academic career in Europe and he appears to have renewed his association years later in Asia (see Chapter 12).

All Jesuit colleges followed a similar curriculum which was formally laid down by the Society in *Ratio Studiorum* (Fitzpatrick 1933). In keeping with these academic arrangements, Strachan's higher education in Paris would have begun with the Trivium. As a rule new students were aged between twelve and fourteen, but Strachan would not have been unusual at sixteen years old. The Trivium normally took five years to complete, the first three of which involved a thorough grounding in Latin and Greek grammar. The fourth year consisted of syntax, with the final year being devoted to the study of rhetoric. On completion, the student was allowed to progress to the higher degree of the Quadrivium, usually at a university college: there the student would choose to study in one or more of four faculties – philosophy (arts), theology, medicine or law. Often serious students would undertake two or more of these disciplines which would require a minimum of eight years' study. The nature of the entries in Strachan's album amicorum indicates that he was a serious and highly regarded student. Many years later Strachan told a friend, Pietro Della Valle, that after he arrived in Paris he made rapid progress in Latin, Greek, philosophy, theology, law and mathematics (Della Valle 1664: vol. 1, 717). This statement implies that he took less than five years to complete the Trivium and, perhaps due to the excellence of his earlier education, may have been excused all of the first three years of grammar. His conversation with Della Valle indicates that, with the exception of the faculty of medicine, he studied in all of the faculties of the Quadrivium during his early years in Paris.

Album Amicorum

Strachan's album amicorum provides clues as to how, when and where he received such a substantial education, but due to the nature of the entries it is not an easy document to interpret and requires unravelling rather than simply a perusal. In owning an album amicorum Strachan was following a fashion among university students which had started in Germany in the middle of the sixteenth century. The German *Stammbuch* was an album in which teachers and fellow students would write greetings and encouraging comments for its owner. Typically they would include verses or sayings in Latin or Greek which were deemed appropriate to the relationship between the inscriber and the owner of the *Stammbuch.* Philipp Melanchthon, the German Protestant reformer, wrote of them:

> These little books certainly have their uses: above all they remind owners of people, and at the same time bring to mind the wise teaching which has been inscribed in them, and they serve as a reminder to the younger students to be industrious in order that the professor may inscribe some kind and commendatory words on parting so that they may always prove themselves brave and virtuous during the remainder of their lives, inspired, even if only through the names of good men, to follow their example. (Nickson 1970: 9–10)

The encomia were to be written on the departure of the student from his college and accumulated throughout his academic career. The medieval custom of *peregrinatio academica*, whereby rather than completing their studies at one college students travelled from university to university to be taught by the best teachers available, was falling into disuse. The establishment of colleges segregated along confessional or national lines in the second half of the sixteenth century had restricted freedom of movement. Students who wished to study at different colleges were presented with practical difficulties related to faith or political allegiances. Nevertheless, gifted students such as George Strachan still studied at a number of different colleges and an album amicorum was a valuable record of their academic success and acted as an introduction to new professors at their next port of call.

The physical form of these books varied. In some cases the student used a published volume by an author he admired and which was already part of his private library. Comments were written in the margins alongside printed passages of particular interest. Sometimes these books were rebound with additional blank pages interleaved with the original text to accommodate the manuscript comments more easily. Most commonly the albums were especially made for the purpose and consisted of blank sheets in a handsome binding. This is the form that George Strachan's album takes. It is quarto-sized (9.5 by 12 inches), has 250 leaves (500 pages) and is bound in fine red leather with an intricate design tooled on the front and back. The centre of the design is a medallion containing the Strachan of Thornton coat of arms showing a stag *passant* with the addition of a star placed between the stag's antlers signifying George's status as the third son of the family (Burke 1884: xxxiii). Strachan's personal seal remains on some of his letters in ASV and again shows the family arms with the stag and star. Above the medallion is an insert with the date 1599 and the words 'Album Amicorum': beneath is 'Georgii Strachani Scoti' ('belonging to George Strachan, Scot'). On the left-hand side of the Thornton arms is the image of the Crucifixion and on the right-hand side that of the Madonna and Child. Veneration of the Virgin Mary became a major point of difference between the Catholic and Protestant Churches. The Anglican Church had removed all but three references to St Mary the Virgin in the Book of Common Prayer introduced in 1549, and discouraged the previous Marian devotions with removal of images. Calvinist and Lutheran Churches were even more thorough in the removal of prayers to Mary. If it were not indicated in any other way, the inclusion of the Madonna and Child on the book's cover would have marked the owner as a Catholic.

Strachan used the album extensively but did not take it with him when he left for the Middle East in 1613. Following his departure it came into the possession of Thomas Chalmers, a graduate of the University of Aberdeen, who had left Scotland to be ordained as a secular priest, settling in Paris in the 1630s. Chalmers' entries in the album date from this time but do not conform to the normal usage of an album amicorum. He used thirty-six leaves to compose a collective obituary which is made up of separate epitaphs for twenty-seven people including his brother David, who had been the principal of the Scots College in Paris. With the exception of Chalmers' work, all other entries in the album relate to Strachan. They fill another ninety-three leaves with the remaining 121 left blank. There is little order to these entries, the writers having used whichever page was convenient. The first inscribed leaf is dated 1603, but the earliest entries that occur later in the album are dated July and August 1599, confirming the embossed date on the book's cover.

These earliest comments were made by professors and fellow students at the small Protestant University of Béarn in Lescar in Navarre and were written on the occasion of Strachan's leaving that institution. There is, however, an entry which throws light on his prior student life. John Barclay, the Scots writer, satirist and, like Strachan, neo-Latin poet, made an entry in the album dated 4 August 1606, in Paris, writing, '*Charae amicitiae foedus cum Giorgio Strachano non nuper initium sed a cunabulis*' thereby claiming that the bond of friendship between him and George Strachan had been formed almost from the cradle ('*a cunabulis*') (Strachan's *Album Amicorum*: 67). Barclay was the son of William Barclay, the jurist, who had been professor of law at the University of Pont-à-Mousson and John had been born there in 1582. It would appear that the young boy had been befriended by Strachan while he was one of his father's students, and years later had renewed the friendship in Paris. The phrase '*a cunabulis*' would indicate that he had made Strachan's acquaintance before adolescence and suggests a date of no later than 1593. If so, George must have completed his Trivium studies and gone to Pont-à-Mousson to study law less than five years after leaving home. The acquaintance with young John Barclay would have formed while Strachan as a senior student acted as his tutor. It was normal practice for Quadrivium students to take on teaching duties in Trivium classes to help with financing their education. It is likely that Strachan had taught young Barclay while still a student of law under the tutelage of the boy's father.

The album gives no information regarding Strachan's life during the six years between 1593 and 1599. Given his impressive academic attainments, he must have continued his studies but his involvement with Jesuits could have caused him difficulties. It is fortunate that he left Claremont in the early 1590s before disaster overtook the college. In 1594 an attempt was made on the life of King Henri IV of France. Henri had inherited the throne as the first Bourbon king and, for political reasons, converted from Protestantism that year. Henri was viewed as a traitor by his Protestant subjects and with suspicion by Catholics. The would-be assassin was a delusional Catholic, Jean Châtel, just nineteen years of age. In the young man's pocket was a confession in which he described how he had been inspired to carry out his attack while undergoing the Jesuit Spiritual Exercises in a *chambre de méditation*. Under questioning, he revealed that he had been educated by the Jesuits at Claremont College. His punishment for *lèse-majesté* was torture and dismemberment but the Parlement de Paris held the Society of Jesus culpable for Châtel's state of mind and ordered an investigation into the college. The college library was found to contain papers critical of the king's accession to the throne and a note in the librarian's handwriting that it was a mistake that Henri had not been killed in the St Bartholomew's Day massacre. Fr Jean Guignard, the unfortunate librarian, was hanged and his corpse burnt at the stake; the college was closed and its property confiscated. Under the circumstances Châtel's former tutors, Fr Guéret and Fr Edmund Hay, who was no longer college principal, were fortunate not to be executed but were sentenced to permanent exile from France. Although the king took the enlightened view that the Society was not involved in the attempt on his life, the Parlement de Paris banished all Jesuits from the city. Exceptionally, they were allowed to remain in other parts of France. Due to this incident the Society was prohibited for more than three decades from having any teaching establishment in Paris (Voltaire 1775: 168–71). George Strachan was not involved in the scandal, as his later friendship with King Henri shows (Chapter 3) and he was able to continue his studies in France, but after these disturbing incidents, matters concerning his family caused George to interrupt his studies and return to Scotland.

Family Difficulties

George's mother died in 1595, and less than two years later his eldest brother, Robert, also died (Balfour 1904: 122). It would have been natural for him to return from France immediately in such circumstances, but, due to the significantly worsening political situation for Catholics in Scotland, he delayed his journey home. Problems had been brewing since 1592 when the country's principal Catholic nobles had been suspected of treason in an episode known as the Spanish Blanks. William Douglas, 10th Earl of Angus, Francis Hay, 9th Earl of Erroll, and George Gordon, 6th Earl later 1st Marquis of Huntly, had been caught in communication with King Philip II of Spain regarding the return of Catholicism to Scotland through support for the Jesuit mission. King James VI did not exact any major penalties from the nobles at the time. There has been speculation that James was not averse to the contacts with the Spanish monarchy (Gribben and Mullan 2009: 138). Later, following his accession to the English throne, James started his own formal negotiations with Spain which led to a peace treaty with Phillip III in 1604. Nevertheless, the Scottish nobles' relationship with the king deteriorated and by 1594 they were in open revolt. They were forced into exile and, in 1597, were only able to return and regain their estates by abjuring Catholicism and declaring themselves to be Calvinist.

George Strachan's father, Alexander, was caught up in the earls' travails. When William Douglas was arraigned before the Privy Council in Edinburgh on 1 August 1597 on a number of charges relating to support for Catholicism, he was allowed his liberty on the provision of a caution of £20,000 Scots. He was also required to deliver his eldest son, his heir, into the custody of William Douglas, 6th Earl of Morton, to be raised in the Protestant faith (Sanderson 2004). The caution needed twenty-three members of the Catholic nobility and gentry collectively in order to guarantee it. One of the guarantors was Alexander Strachan of Thornton whose daughter-in-law, Robert's widow Sarah, was the earl's daughter (Masson 1884: vol. 5, 745). Clearly this placed the Strachan family in a very dangerous position. Not only would they be financially exposed if the earl was found to be in default but members of the family were vulnerable to accusations of breach of the Penal Laws from Protestant zealots. Later, the financial risk proved to be real. On 12 February 1607 the king's advocate, Sir Thomas Hamilton of Monckton, brought a case against the Earl of Angus for aiding the Jesuit Johne McGhie (Masson 1884: vol. 7, 317–18). One of the strictures against Douglas in the original hearing of 1597 was that he 'shall not reset Jesuits, seminary priests or excommunicat papists'. Twenty-one of the original twenty-three guarantors were cited in the second court case. George Strachan's father had died several years before and been succeeded by Alexander, Robert's son, who was called to court in his grandfather's place. The case was dismissed when the defendants swore that they did not know that McGhie had been excommunicated.

During this period the whole Catholic community of nobility and gentry came under intense pressure and many followed the example of the three earls and subscribed to the Calvinist Confession of Faith. But even after conversion these previously Catholic families were open to attack from Protestant vigilantes who sought personal benefit by calling into question the sincerity of their conversion. The Strachan family was not exempt from these pressures and took protective action. George's elder brother, John, not only converted but became the Calvinist minister of the local parish of Kincardine. In doing so John seems to have been making a virtue of necessity. The parish living was in the gift of the Strachan family as they were the principal local landowners. John's appointment as minister not only blunted Calvinist antagonism towards the family but also solved the problem of providing its second son with financial support and a suitable career.

In itself John's apostasy did not resolve the family's predicament. Although Catholics already in possession of property could retain their estates, they faced fines levied by the Kirk for non-attendance at church: in addition the Penal Laws prohibited Catholics from inheriting property. The nearest Protestant relative could lay claim to the estate of any deceased Catholic irrespective of the existence of closer heirs. Often relatives refused to take advantage of this provision but the law was to become increasingly arduous when even distant relations could dispossess the rightful heir. Families were forced to take hard decisions to retain their estates. A number resolved the issue by means of the head of the house 'converting' to Calvinism while the remainder held to the Catholic faith. A notable case involved the Urquharts of Meldrum whose eldest son, John, on succeeding his father converted to Calvinism while his two brothers were ordained as Jesuits (Forbes-Leith 1909: 186). By these means the Kirk and State tried to eradicate loyalty to the 'Old Faith' that many, particularly in the north-east of Scotland, still professed.

When George's eldest brother, Robert, died his son, Alexander, who was still a minor at the time, became his grandfather's heir. While the old man lived, the family's possession of the estate was secure but he was ailing and was to die about three years after George's visit home. The family realised that it could not be long before young Alexander took his place. His mother, Sarah, as his guardian, assumed an important position in the family's affairs. These were the prevailing circumstances when George did eventually return to Scotland to see his family. The visit is alluded to in a series of letters dated 1602 sent to the Roman authorities from Jesuits in Scotland (Chapter 2). The exact date of his return is uncertain but was between 1597 and 1599. The family reunion must have been strained. During George's visit the conversations, which he held with the new matriarch and the rest of the immediate family, could only have raised concerns for everyone. He made them aware of his respect for the Society of Jesus and that he was considering becoming a Jesuit. Later it was erroneously reported that he had joined the Society of Jesus (Pattison 1892: 215). This would have been a logical outcome of his life thus far. Prior to his reaching the age of maturity, the greater part of his education had been provided by Jesuit teachers. As an intelligent student from a good family, he would have been welcomed into the Society and his teachers would have taken pains to convince him of the attractions of joining their elite order. The family, especially his brother John, the Calvinist minister, could only have viewed the prospect with alarm. Alexander's mother would have been aware that the safest way to protect their interests was for the rest of the family, especially her son, to follow the example of her father, the Earl of Angus, and convert to Calvinism. The sincerity of their conversion, which may have happened before George's visit, would have been viewed with suspicion by the Kirk if a close member of the family had become a Jesuit. From later correspondence it is clear that family relations were difficult, but George appears to have been sensitive to the family's predicament, as was shown by his actions on his return to the continent. During his visit George must have gained some additional financial support, perhaps a legacy from his mother, since on his return to France in 1598 or early 1599 he was able to indulge in the expensive but non-essential purchase of his album amicorum. The bequest, if that is what it was, would have strained the family coffers and proved to be the last financial support that George was to receive from them.

When his father died in 1600–1, the extent of the Strachan family's money difficulties became clear. Immediately on succeeding to the title young Alexander was pursued for his grandfather's debts. The creditor, Alexander Elphingston, the son of the Master of Elphingston, tried to gain the whole estate by charging the deceased as a rebel and the new Strachan of Thornton as complicit. Elphingston claimed at a hearing before the Privy Council on 12 December 1601 that he had sent a sheriff's officer to take possession of the house in lieu of a debt, and that young Alexander had refused to recognise the authority of the writ. The Privy Council refused to come to a judgment regarding the accusation of rebellion on the grounds that Alexander was a minor. Another hearing was required to settle the matter of the debt (Masson 1884: vol. 6, 710). The record of the case in the register erroneously gives Alexander's name as John and describes him as the son of the deceased Alexander. This would mean that the minister of Kincardine was the defendant. However, the case must refer to Alexander since it describes him as a minor and heir to Alexander Strachan of Thornton. The significance of finding Alexander guilty of rebellion was that all of the individual's property would be confiscated. In those circumstances Elphingston, as creditor, would have been in a privileged position to acquire the estate on advantageous terms. At the hearing on 24 December, Alexander explained that he had offered to hand the house over to the officer but the officer declined, saying that he had other business to attend to first. The officer and Elphingston were in collusion to dispossess Strachan completely. The accusation of being a rebel only required that the individual defy the authority of the king (in the form of the court writ). If he had refused to hand over the house, he would have fallen foul of the law and been declared a rebel. Alexander's version of events was accepted and the council threw out the case but the debt still required to be paid. The amount was 500 merks and the court records show that a relative, Roger Strachane (sic) of Glethknow, stood surety for his young kinsman (Masson 1884: vol. 6, 326). It would appear that, at his death, Alexander Strachan had left the family's financial affairs in serious disarray. George returned to France before these events played out but it is clear from his later behaviour that he understood that in the future he would have to rely on his own efforts to make his way in life.

CHAPTER TWO

Exile

Peregrinations

In 1593 when Strachan reached the age of twenty-one, he assumed financial responsibility for himself. It would appear that his family continued to provide some support, but there is sufficient evidence to suggest that obtaining an income became increasingly important to him. By that time he had spent five years abroad as a student and was showing a desire to continue his life as a serious scholar. For someone of his limited means, this presented a problem. While completing the Quadrivium, he could support himself in part by teaching younger students. As a gifted scholar, he could obtain a teaching post at a university or college, but such a career would never have been able to provide sufficient income for someone of his social status. His good friend, Thomas Dempster of Muiresk, similarly a younger son of Catholic gentry in the north of Scotland, was unable to support himself while simultaneously holding professorships at four different colleges at the University of Paris. He needed the sponsorship of a wealthy benefactor; until he obtained one he was unable to marry or support a family (Du Toit 2004).

Faced with similar problems while still in his early twenties, Strachan had given consideration to a life in the Church. The views of his family regarding the difficulties this would cause them must have given him some concern. Following his visit home in the late 1590s, he did not abandon all thoughts he had on the matter, but merely refrained from taking a decision. For over a decade the Jesuits continued to hope that he would join the Society. On his return to France in 1598–9, his primary interests still lay in academic study, but it is clear that he had financial problems. Whatever limited money Strachan had received from his family on his visit could not have lasted long. They were in no position to be generous.

Despite the relative bleakness of his prospects, he showed a certain confidence in his financial position since it was on his return to France that Strachan bought his album amicorum. The cost was roughly seventy merks. Parisian booksellers of the time charged two denarii per sheet of paper (Proot 2018: 199–200). The 250 folio sheets unbound would have cost the equivalent of £2 sterling. The leather binding with the gold leaf tooling would have doubled the cost. The expense incurred in its purchase may not appear excessive, but it represented a non-essential purchase for a young man of limited means whose family was beset by creditors. However, it is fortunate for researchers that he chose to spend his money in this way. The album helps identify the developments in Strachan's career through the light it sheds on his movements over the following decade. The book is bound in French leather and it is likely that Strachan purchased it in Paris before he left for Navarre early in 1599.

The first entries dated July and August of that year were made by his professors and fellow students at the University of Lescar in the province of Béarn (Strachan's *Album Amicorum*: 27). It is clear from their remarks that Strachan had been residing at the university for some months. Lescar was a Calvinist institution with an excellent school of medicine and normally only accepted students of reformed faith. It would have been impossible for Strachan to have been allowed to enrol while openly declaring his Catholicism. Admission to the college would have required, on Strachan's part, some expression of adherence to the reformed religion. Any action of that nature must have been insincere since he later showed that he had not wavered in his adherence to Catholicism. It was relatively common for scholars to pretend to change their religious adherence in order to gain admittance to colleges for education or employment purposes. Often the college authorities colluded in such deceptions when the applicant was one of outstanding ability. This must have been the case with Strachan in Lescar. The college principal could not have failed to see the image of the Madonna and Child on the cover of the album when he made his entry in it praising Strachan's merits as a scholar.

Thomas Dempster had a similar experience when he joined the Protestant College of La Rochelle after declaring himself a Calvinist. Despite the improbability of conversion by such a renowned Catholic scholar, the college authorities did not question his statement and thereby obtained the services of this distinguished humanist (McInally 2012a: 137).

Similarly, Protestant scholars claimed conversion to Catholicism. Thomas Reid and Thomas Seget, both Scots Calvinists, attended the Catholic Scots College in Louvain in 1596 in order to study under Justus Lipsius, the renowned philosopher and founder of the Neo-Stoicism school (McInally 2012a: 137). Strachan's attendance at the college in Béarn can be viewed in a similar light – an expediency necessary to further his academic career. At the time there appears to have been a strong Scottish contingent at the university. James Fleming and Gilbert Burnet were professors and each made a full-page entry in the album along with one of the students, John Hamilton (Strachan's *Album Amicorum*: 23). He was the son of Claud, 1st Lord Paisley; in 1606 Hamilton was created 1st Earl of Abercorn. In his album entry, he describes Strachan as a loyal friend. Both were Catholic and appear to have struck up their friendship while conspiring to persuade the university authorities of their supposed conversion to Calvinism. No matter the degree of dissemblance involved in his declaration of faith, Strachan's public ambiguity concerning his confessional allegiance indicates that, perhaps sensitive to his family's predicament, he had begun to distance himself from the Jesuits.

Shortly after leaving Lescar he spent brief periods at the Catholic universities of Toulouse and Bordeaux before, in 1600, visiting Montauban which at the time was an independent republic and centre of Huguenot power in the south of France. The album records in the same year that he also studied at the Protestant University of Montpellier. Like Lescar, Montpellier had a renowned medical school. In each case Strachan did not stay long, never more than a few months: a pattern that would indicate that he was not engaging in a prolonged course of study but was looking for employment or a patron. This would also explain his visit to Montauban where there was no educational establishment of any significance but where leading Huguenots would have been able to provide him with introductions to the prestigious University of Montpellier and could have sponsored him if they had wished. There are over two dozen entries in the album over this period, made by prominent Scotsmen studying and teaching abroad (McRoberts 1952: 111). Whatever hopes Strachan had, regarding patronage or lucrative employment in these establishments, went unfulfilled.

Bishop William Chisholm

A year later Strachan was in Carpentras in the Venaissin in the Rhone Valley. The Venaissin, along with neighbouring Avignon, was an enclave of papal territory in France. Carpentras was its principal population centre and Vaison was the seat of the bishop. He had gone there to meet William Chisholm, a Scot who had 'inherited' the bishopric from his uncle of the same name (Dilworth 2004). The elder Chisholm had been Bishop of Dunblane in Scotland at the time of the Reformation and had been a vigorous opponent of John Knox. After he lost his Scottish diocese, Pope Pius V compensated Chisholm with the See of Vaison. When he retired to a monastery in 1584 his nephew succeeded him and continued with his uncle's involvement in the affairs of Scotland.

As part of his efforts to influence developments at home, the younger Chisholm published a polemic against Calvinism, *Examen Confessionis Fidei Calvinianae quam Scotis subscribendum proponent.* Ostensibly the reason for Strachan's visit was to present the bishop with a Latin poem that he had composed in praise of this work, but he was also hoping for some preferment (Leask 1910: 338–40). Strachan's involvement with Chisholm's affairs lasted for some time but provided little financial reward. The bishop added some complimentary remarks to his album amicorum and appears to have given Strachan a commission. At the time Chisholm, at the behest of King James VI of Scotland, was being considered by Pope Clement VIII for elevation to the College of Cardinals. The likelihood of James succeeding to the throne of England had increased the importance of Scotland in the eyes of European rulers. James had been courted by a series of popes in the hope that he would convert to Catholicism in exchange for the support of European Catholic powers in his claim on England. The king had encouraged these diplomatic approaches, and Scots Catholics such as Chisholm took advantage of the changed political fortunes to gain advancement in the Church. Earlier Chisholm had petitioned the pope to set up a Scots College in Rome using the medieval Scots pilgrims' hospice in the city. As part of the jubilee celebrations of 1600, Pope Clement had consented to the establishment of the Pontifical Scots College in the Eternal City but that was the extent of whatever influence Chisholm had with the pope. At the time of Strachan's meeting with Chisholm in 1601, King James was distancing himself from the papal overtures; English courtiers had started discussions with him regarding his accession on the awaited death of Elizabeth (Black 1959: 495). James was about to achieve his objective without the need of Catholic support. The cardinal's hat never materialised for Chisholm but he was reluctant to accept the change in his fortunes and appears to have entrusted Strachan with a letter for King James. The attempt to revive the latter's support for his appointment to the Curia had to be carried out discreetly if it were to have any chance of success. In 1601 Strachan made a brief secret visit to Scotland.

The album amicorum sheds no light on this incident. There is a lacuna in the book between Chisholm's entry in 1601 and one by Alexander Scot in Rome the following year. The evidence for Strachan's trip comes from a letter written by a Jesuit in Scotland. The Jesuit missionaries were well informed of developments at the Scottish court and kept the pope informed on the changes in the king's attitude to Catholicism. Robert Abercrombie was personal confessor to James' wife, Queen Anne, having recently received her into the Catholic Church. His was a privileged position that allowed him to gain inside knowledge of diplomatic developments and court gossip. Alexander MacQuhirrie, one of Abercrombie's colleagues, compiled a report specifically for the pope's attention on the state of Catholicism in the country. In his letter MacQuhirrie informed the pope that the courier, a gentleman named Strachan, would be able to elaborate on his report since he was fully acquainted with the situation in Scotland. It appears that Strachan was directed by Bishop Chisholm to contact Robert Abercrombie at the queen's court in order to arrange the secure and secret delivery of his confidential letter to the king. It is likely that Chisholm had used this route with his earlier communications. Abercrombie may have been able to arrange that Strachan deliver the letter and receive the king's answer in person. The king's answer would have disappointed Chisholm. After delivery of Chisholm's letter to James, Abercrombie entrusted Strachan with MacQuhirrie's report for the pope.

Strachan's presence in Scotland was brief and its purpose did not allow him to contact his family. This must have been the case since the difficulties which arose with his family in the summer of 1602 (see below) would have materialised during this visit and possibly have had a more severe outcome for him. On his return to the Continent, he went first to Chisholm in Venaissin to give an account of his visit but quickly travelled on to Rome to deliver MacQuhirrie's letter in person to Pope Clement (Forbes Leith 1885: 269–74). Since he was on Jesuit business, General Aquaviva would have vetted Strachan personally for such an important meeting. For Strachan the opportunity of an interview with the pope and the Jesuit general would have raised his hopes of patronage within the Church. Unfortunately for Strachan, MacQuhirrie had made clear in his report that there was little likelihood of the conversion of King James to Catholicism. Papal hopes in this regard had waxed and waned over the four decades of the king's life. However, by 1601 it was clear that the king's interests lay in remaining Protestant. Strachan was, therefore, the bearer of unwelcome although not unexpected news. There was no reward of position or pension for his services.

Scots College, Rome

When Strachan arrived in Rome the pope's new Scots College had just started to admit students and although he had already gained a higher level of education than the college was able to offer, Strachan enrolled as one of its first students. This was a purely opportunistic move on his part, since the college provided its students with bed and board which allowed him to remain in Rome and look for career opportunities. Again, the album amicorum throws light on Strachan's activities. It contains an entry dated 1602 by a fellow countryman, Alexander Scot, in which Scot speaks of their close friendship and of his respect for Strachan's scholarship. Scot was a Catholic humanist who, from 1594, had been the principal regent at the University of Carpentras and was on a visit to Rome to promote his work. The second edition of his lexicon on the works of Cicero had been newly published. It included a Latin epigram in praise of Scot's learning and piety written by Strachan (Dellavida 1956: 4–5). The two Scotsmen were simply renewing their friendship in Rome. They were acquainted from Strachan's visit to Bishop Chisholm in Carpentras and possibly earlier when studying in Paris. Strachan also made the acquaintance of another travelling scholar, the mathematician Marino Ghetaldi, who like Scot was in the city to arrange publication of his work (Durkan 1971: 12). Ghetaldi was a Venetian citizen from Ragusa (present-day Dubrovnik) and a friend of Galileo. As well as both being mathematicians, Strachan and the Venetian had much in common: they were around the same age – Ghetaldi was about four years older – both were younger sons of minor nobility and they had had a wide-ranging humanist education much of which had been acquired in France.

Strachan stayed in Rome meeting more expatriate Scots and gaining valuable contacts in academe and the Church (McRoberts 1952: 113). Despite the disappointment of Bishop Chisholm's failure to be promoted to cardinal, Strachan seems to have believed that his best hopes for advancement lay in patronage within the Church. He made himself known to the cardinal protector of Scotland, Camilo Borghese, who, shortly afterwards in 1605, was created pope as Paul V. Later, in a poem celebrating Pope Paul's election, Strachan described Borghese as his patron and thanked him for the many favours which he had received from the cardinal when he was in Rome (Dellavida 1956: 19–20). One of those favours was likely to have been permission to enrol at the Scots College. He was thirty years of age – much older than the young men in their teens for whom the college had been established. The level of education offered was such that he should have been a professor rather than a student and it is doubtful that he had any means of paying for his board. The college principal, Mgr. Bernardino Paolini (Durkan 1971: 12), who was a papal appointee, would have needed dispensation from someone of the status of the cardinal protector before allowing Strachan admittance. Also it is likely that it was through Borghese that Strachan became acquainted with Cardinal Federico Borromeo, Archbishop of Milan (Durkan 1971: 12). Borromeo, cousin to St Carlo Borromeo, was of a similar age to Strachan and like him a scholar and writer who could provide him with further introductions in the Church. This may have been a reason why, after leaving Rome, Strachan travelled through the province of Ticino and to Pavia in northern Italy to meet up again with Federico and a number of other senior prelates in the region. However, it was not why he left Rome.

Letter from Home

Claudio Aquaviva had use for Strachan as a courier to the Jesuit missionaries in Scotland. At some point prior to his arrival in Rome in 1602, Strachan had received a letter from home which contained news of his father's death. George's nephew had inherited his grandfather's title and lands as Sir Alexander the 13th Strachan of Thornton. He was still a minor and, as his guardian, his mother, the now dowager Lady Strachan, took control of the affairs of the family. In law minors were not required to declare any religious affiliation and were immune from sequestration of their property due to recusancy. The same protection applied to guardians since their control of the estate was temporary, limited to the duration of the minority. Alexander would not have reached the age of majority for at least another eight years but Lady Strachan was worried about debts and legal pressures. Her concern was to protect her son's inheritance. Her father, the Earl of Angus, had already ascribed to the Calvinist Confession of Faith as a condition of his return from exile, and it can be assumed that the advice given by her brother-in-law, John, the minister of Kincardine, would have been to follow her father's example. She wrote to George informing him of his father's death and, in terms of conciliation and friendship, discussed the family's future. It is not clear when the Strachan family decided that they should become Calvinist but it may have been after Lady Strachan wrote her letter. By the following year at the latest, they had converted; on his return home in 1602 George was faced with a *fait accompli.* Robert Abercrombie, the prefect of the Jesuit missionaries in Scotland, later described the contents of Lady Strachan's letter as being full of expressions of familial affection, but in reality it was a treacherous invitation to George to return home, where he was put under pressure to convert to Calvinism (ARSI, *Anglia,* 42: F167r–v).

This seems a distorted interpretation of what was probably a genuine expression of family grief at the death of Sir Alexander and genuine concern for the future of the family fortunes. It is likely that the family did not expect George to return immediately, if at all. Lady Strachan may not have requested his presence, since he waited at least a year after his father's death before setting out for Scotland. It is difficult to see what advantage could have accrued to the Strachan family from the return of the impecunious George. They were already facing debts which were to force them to borrow from a kinsman in order to avoid court sanctions. As it proved, his return only added to their problems, but George decided to use the news of his father's death as a reason to return to Scotland. General Aquaviva saw Strachan as a useful courier and following his meeting with the pope he asked him to take a letter to Abercrombie. Strachan agreed and the decision was to change the course of his life.

Entries in his album amicorum show that he set out from Rome in the early summer of 1602 and took a circuitous route. He first visited the Italian alpine province of Ticino and then the city of Pavia. Afterwards he went to Vaison and Carpentras possibly to report to the bishop on what he had learned in Rome of the progress, or lack of it, on the bishop's elevation to the Curia. More likely he offered to carry any correspondence that the bishop wanted delivered in Scotland. Again his stay was brief, and he left to meet up with friends in Paris before moving on to arrive in Scotland some time between June and September. Immediately he was in difficulties with the Kirk authorities. In a letter dated 20 September 1602, the Jesuit missionary Andrew Stinson told of Strachan's arrival. The letter was addressed to Aquaviva's secretary, Fr George Duras. Stinson wrote that the general's letter which Strachan had delivered had given them all great comfort (ARSI, *Anglia,* 42: F157). He also informed Fr Duras that Strachan had been deceived by his family. He petitioned the secretary on Strachan's behalf, saying that he was deserving of appointment to a distinguished position when he returned to Rome.

Stinson's comments were expanded upon by Abercrombie in his letter to Aquaviva dated 13 December 1602 (ARSI, *Anglia,* 42, F167r–v). Abercrombie's version of events was that the Strachan family had converted to Calvinism, and had deliberately recalled George to Scotland in order to persuade him to follow their example. George had refused and they reported him to the Kirk authorities who accused him of the crimes of apostasy, of being a traitor to his own country and a spy and agent of the pope. Abercrombie wrote that the Strachan family explained George's behaviour by saying that he had been seduced by gold and riches.

On Trial

Clearly Strachan was in very serious danger. The accusation of treason could not be decided by the Kirk Session; only the Privy Council could rule on such a matter. In his account Abercrombie stated that the king was prejudiced against Strachan's accusers from the outset and sympathetic to the accused (ARSI, *Anglia*, 42: F167r–v). This is almost certainly true as the court was never favourably disposed to Presbyterians, since they refused to recognise the king's supremacy on spiritual matters within the Church of Scotland. Although they were Calvinists, the Strachan family did not belong to the Presbyterian faction of the Kirk. George's brother John had gained his parish through family patronage rather than appointment by the presbytery, which indicates that they belonged to the Episcopal tendency in the Calvinist Church. However, in Edinburgh where the court and Privy Council met, Presbyterians were dominant. Strachan was unfortunate in that, at the time of the court hearing, an assembly meeting of the Church of Scotland was being held and anti-papist sentiment among the majority ensured that extreme interest was shown in Strachan's case. Rather than allow the Privy Council to deal with the matter the king decided to hear the case himself. He may have been influenced in this by Strachan's visit to Scotland the year before. He would not have wanted the Privy Council to listen to testimony which might have exposed the correspondence between himself and Bishop Chisholm and his former dealings with the pope. Whatever King James may have felt regarding possibly showing leniency to Strachan, he would not have wanted a confrontation with the Kirk assembly. Nevertheless, the court would have viewed Strachan as a member of their privileged society of nobles and gentry. Furthermore, James considered himself a distinguished scholar having published on a number of subjects, and would have had a personal empathy with someone he saw as a fellow humanist, a point which Strachan would have taken advantage of in his defense. Abercrombie wrote in his letter that at his trial George was able to win over James on the key points of the accusations. The charge of apostasy was thrown out on the grounds that Strachan had never espoused Calvinism or subscribed to the Confession of Faith. The evidence that he was a papal spy was based on his having been in Rome. Strachan argued that as a scholar he had occasion to travel to many parts of Europe, and his visit to Rome was, therefore, unremarkable. The charge of treason was based on the accusation of spying and both were dismissed by the court.

George Strachan had faced serious accusations, but Abercrombie's version of events tells only part of the story and in a number of respects he is mistaken. While describing the Strachan family's treachery towards George, he claimed that all had turned against him, including his mother.

Since George's mother had died eight years earlier, he seems to have confused Robert's widow, dowager Lady Strachan, with the earlier Lady Strachan. This inaccurate and to a certain extent biased account was given by Abercrombie in order to present Strachan as someone who had suffered for his faith. Like Stinson he felt that Strachan should be given a sympathetic reception in Rome.

He was correct in writing that the family had converted to Calvinism. The family's straitened financial affairs made payment of the Kirk's fines for non-conformity impossible. It is likely that Lady Strachan had subscribed to the Calvinist Confession of Faith on behalf of the whole family thereby protecting everyone from financial penalties. Prior to his visit, George may not have known of these developments or his nephew's appearance before the Privy Council the previous December to answer for his grandfather's debts. George's return to the Mearns presented the family with the dilemma of either protecting him at great cost to themselves or persuading him to cooperate with their change of faith. Despite their familial bonds, they could not protect him without risking financial ruin. In their eyes the solution appeared to be that he should dissemble regarding his Catholicism but, as Stinson and Abercrombie reported, he would not agree to such an act. Through his brother John, his presence at Thornton Castle would have been known to the local ministers almost as soon as he arrived, and they would have lost no time in confronting him. His refusal to conform to Calvinism allowed them to accuse him of apostasy. The fact that he had been in Rome and had been given his travelling expenses clearly made him a papal agent and spy in their eyes. Trial by the Privy Council was required, since if proven the charges carried the death penalty. Abercrombie's statement that it was King James' wish to hear the case himself is almost certainly true, but his assertion of a successful outcome for Strachan is only partly correct. Strachan had argued that there were no grounds on which to convict him on any of the charges, especially that of treason, and King James and the aristocratic members of his court who heard the case were happy to accept his robust rebuttal of the Presbyterians' accusations. George Strachan may have reminded the court that he was a direct descendant of King James I and this would have played a part in influencing the king, his judge, to dismiss the case. In addition James would have sympathised with Strachan due to his status as a scholar.

Although the court dismissed the charges of apostasy, espionage and treason, Strachan was a self-confessed practising Catholic. In light of this, the law required that he be sentenced to permanent exile. Such a sentence also relieved the king of running the risk that Strachan, at some point in the future, might disclose their earlier dealings. Perhaps, it was in an attempt to retain Strachan's silence that, exceptionally, the king gave him a three-month suspension of execution of the sentence in which he was free to move around the country concluding his business and bidding his family and friends goodbye.

During this period of grace he travelled around the east and north-east of Scotland, meeting with more of the Jesuit missionaries and visiting local Catholic families (Strachan's *Album Amicorum*: 20, 80, 38). It was fortunate that earlier Strachan had been discreet in his contacts with the Jesuits and that they had made no contributions to his album amicorum. Aiding Catholic priests was a crime which carried heavy penalties. The entries in his book during his stay in Scotland were made after his trial, and only by members of the gentry expressing their friendship and admiration for his scholarship. An exception was the contribution of Robert Blackwood, who had been a presbyter of Brechin Cathedral at the time of the Reformation and had remained a Catholic priest in hiding for over forty years. If found out, this could have compromised Strachan but in his entry Blackwood described himself as his uncle. He also expressed his gratitude for the kindness that George had shown him, stating that it was more than he had received from other family members during his long life – he was over eighty years of age (Strachan's *Album Amicorum*: 79).

On the surface this behaviour fitted with the image which Strachan had presented to the court: that of a scholar of distinction within the European humanist community who was visiting Scotland after an absence of a number of years for the purpose of seeing his family and renewing his acquaintance with Scottish gentlemen whom he had met while travelling on the Continent. The king had been happy to agree with Strachan's version, and no doubt enjoyed depriving the Presbyterians of their victim, but the sentence of permanent exile was to create particular difficulties for Strachan. He was in his early thirties with the greater part of his adult life ahead of him. He had been educated to a higher degree than most of his contemporaries and, in addition to the loss of ties of family and friendship that the sentence of exile involved, he was deprived of opportunities to earn a living in his own country at a level that would have allowed him to maintain his social standing. His punishment was to prove even more onerous when the following year James became king of England as well as Scotland, and Strachan's exile applied throughout all the islands of Britain. At the age of thirty, Strachan had still not acquired the financial means of supporting himself in the academic life that he was intent on pursuing, and was to face a continuing struggle in his attempts to do so.

CHAPTER THREE

The Humanist Scholar

Jesuit Leanings

George Strachan had gained one benefit from his visit to Scotland: he had earned the gratitude and sympathies of the Scottish Jesuits. In their eyes he had suffered for his faith while acting to fulfil the commission given to him by Aquaviva. In their letters both Stinson and Abercrombie made special appeals to the general that Strachan 'was deserving of appointment to a distinguished position when he returned to Rome'. Their intention in doing so was to help him gain a reward for his efforts on behalf of the Society, and provide compensation for the sacrifices he had made for his faith. They had come to think of him as one of them and Strachan had led them to believe that he intended joining their Society. Abercrombie said as much in his letter, adding that Strachan would be returning to Rome shortly and would give the general a full report on affairs in Scotland in person.

He showed his complete faith in him by entrusting him with another matter. As part of their mission the Jesuits actively recruited students in Scotland to study as seminarians in the Catholic colleges abroad. Normally it was a missionary who escorted these young men to a Jesuit-run college on the Continent. Abercrombie asked Strachan to perform this task, which demonstrates the degree to which he viewed George as part of their Society. By 1602, in addition to Braunsberg, there were two colleges which were exclusively for Scots: one at Douai in the Spanish Netherlands and the Pontifical Scots College in Rome (McInally 2012b: 6–61). The following year a further college was opened in Paris that, due to the Jesuits' expulsion from the city in the previous decade, was run by secular priests (Chapter 1).

At the time of Strachan's visit to Scotland, twelve-year-old Patrick Seton, son of the Laird of Parbroath in Fife, and nephew of Lord Seton, James VI's late chancellor, was waiting to be taken abroad to study. His widowed mother had asked Abercrombie for his help in this. Patrick was the sixth of her nine sons. Her five eldest had reached manhood and were no longer Lady Seton's responsibility but the younger ones required a higher education which she could not afford. Her wish was that Patrick should travel to a Catholic college on the Continent as his uncle had done in the 1570s. The three younger boys were still receiving their elementary education at home but Abercrombie wrote telling Aquaviva that it was his intention, when the time came, to send them to be educated at Braunsberg. However, he had decided to send Patrick to the Scots College in Rome on Strachan's recommendation. George had given him a good report of the newly opened college and Abercrombie was relying on the general's support to gain admission for young Seton without charge to the family. Strachan was prepared to act as guide for Patrick and it is possible that he was known to the family. On his return to Paris, an Alexander Setone (sic) and Meldrum wrote a message in Strachan's album amicorum expressing a bond of friendship between himself and George. The message may have been an expression of gratitude for Strachan's care of young Patrick, but it is more likely that the connection pre-existed. Strachan's friendship with a kinsman would have reassured Lady Seton regarding his stewardship of her son (Strachan's *Album Amicorum*: 48).

Patrick Seton and Strachan were joined by another young man whom George described as his nephew. This travelling companion is not named. Although it is possible that he was a son of his brother John, it is more likely that he was the son of the eldest of his three sisters, Magdalene, and her first husband, George Symmer of Balzordie. Magdalene had been widowed, and in 1601 remarried to a William Rait of Hallgreen who, like her brother John, was a Calvinist minister. The boy would have been about twelve years of age, and her new husband may have decided that it was better that the youth should not be intimate with his new family. In those circumstances sending him abroad to continue his education would not be surprising, although a Calvinist education would have been required. The most popular destination for Scottish Calvinist students was Geneva. By acting as guide to his nephew, George demonstrated that, despite his recent experience, his familial bonds remained intact. Under these circumstances, although there is no evidence of later contact, it is difficult to believe that he would not have retained an interest in his family's affairs.

Strachan and his small party sailed from the river Tay in January 1603 on the first stage of their journey to Rome. This can be deduced from an entry in his album amicorum dated 2 January 1603, which was made by Walter Rollok of Craigie. Appended to this entry is a place name, Wiest, which has been read, in my view mistakenly, by Dellavida and others as somewhere on the Continent – there are a number of places called Wiest in Germany, Austria and elsewhere. The suggestion has also been made that it refers to a small town in Flanders. The entry makes more sense if Strachan and his party were staying with Rollok at his home in Craigie, which was close to Dundee, while they waited for a ship to take them to the Continent. Wiest or Wicst (the handwriting is unclear) would then be very local to Dundee (Strachan's *Album Amicorum*: 70).

The journey nearly ended in disaster when their ship was wrecked off the coast of France. Later Strachan was to describe in his memorial poem 'Lacrymae' (see below) how they had narrowly escaped death and how he had struggled ashore with Patrick in his arms and his nephew clinging to him. Strachan's album amicorum bears witness to this dramatic episode in his life. The first 200 pages show obvious signs of water damage. Much as he valued the book, it appears that he was unable to protect it completely while rescuing himself and his two charges. According to entries in the album, the party spent some months in Paris recovering from their misadventure before leaving for Rome (Strachan's *Album Amicorum*: 27). Strachan makes no further mention of his nephew. The boy may have remained in Paris and enrolled at a college in that city. However, in 'Lacrymae' Strachan described his onward journey as being along the Loire and Rhône and then an arduous crossing of the Alps. It is difficult to explain why they should have made such a difficult diversion from the easier route to the south of France unless the purpose was to take his nephew to Geneva to be enrolled there in a Calvinist college. Strachan's silence on the matter is understandable given the circumstances of their confessional differences. Strachan and Seton then travelled south by the pilgrimage route through northern Italy to Rome. Their journey from Scotland had lasted nearly a year. They arrived in Rome no later than 6 January 1604 (Strachan's *Album Amicorum*: 97) where Patrick Seton enrolled in the Pontifical Scots College (Anderson 1906: 101).

A Long Journey in Pleasant Company

On return to Rome Strachan met with General Aquaviva but any discussion between the two has not been recorded. The general would have been expecting a report of the visit to Scotland, since Abercrombie's letter of December 1602 advising him of Strachan's situation had been delivered earlier. Abercrombie must have considered it too risky to entrust the letter to Strachan, while he was under sentence of exile and subject to the scrutiny of the Scottish authorities. No doubt Aquaviva would have been interested to hear Strachan's account of affairs in Scotland, but the letter would have led him to expect the Scotsman to express a desire to enter the Society of Jesus. If that had been Strachan's original intention, he did not act on it. Whatever the two men discussed, it became clear to Strachan that he could not expect substantial preferment from the general.

His stay in Rome was short-lived and by 7 March he was in Venice. He had travelled in the company of Patrick Stichel, a Jesuit, who was on his way to Scotland to join Abercrombie as a missionary. Stichel's entry in Strachan's album amicorum, made at their parting in Venice, stated that he had had '*longae pereginationis jocunda societas*' (a long journey in pleasant company), an indication of the enjoyable nature of Strachan's company and the ease with which he made friends (Strachan's *Album Amicorum*: 191). According to one account of Strachan's life, while in Venice at this time 'he taught with great applause' (Dellavida 1956: 13) but the brevity of his stay – he soon left to travel to Paris – suggests that he had not gone there to take up a teaching post. It is possible that General Aquaviva had asked him to undertake another mission on behalf of the Society.

His visit coincided with a period when the relationship between the

Republic of Venice and the Papacy had deteriorated to a dangerous point due to disputes over property rights. For over 100 years the republic and the papacy had disputed territory on the borders of the Veneto and the Papal States. On occasion they had gone to war, and in the early sixteenth century the pope had gone as far as excommunicating the entire republic. For centuries the *signoria*, through necessity, had strictly controlled use of the limited land available on the islands in the lagoon. At the end of the sixteenth century a new doge, Marino Grimani, showed his antagonism to Church authority by extending restrictions, specific to the Church, to Terra Firma, the Venetian territory on the mainland. In 1602 he forbade religious orders from repossessing land that they had rented to lay persons. This effectively dispossessed the Church of much of its property. In the following year the doge enacted a law prohibiting the building of churches anywhere without express permission of the *signoria*. The papacy reacted to these restrictions with increasing indignation. When Strachan arrived in Venice, another curtailment of Church property rights had just been enacted by the state. Prior to the Reformation, the Church had built up enormous property holdings throughout Europe. Venice was determined that the Church – effectively the papacy – would not accumulate large landholdings in the republic. The new Venetian law stipulated that any pious donations of land to the Church could be held for a maximum of two years, after which the property had to be sold to a Venetian citizen with the proceeds of the sale being kept by the Church. The laws applied to all branches of the Church, but much of the Venetians' hostility was directed at the Society of Jesus, since it was seen as the most vigorous supporter of papal authority.

Strachan's journey to Venice at such a febrile time in the relationship between the republic and the papacy may not have been coincidental. If he had been sent by Aquaviva, he may have been delivering a message to the Jesuit principal at the Society's headquarters in Venice, their church and school of Santa Maria dell'Umilità. Aquaviva needed to give him instructions and advice on how to react to the developing situation. The following year, matters came to a head. Pope Clement VIII placed the entire republic under an interdict forbidding all ordained priests from performing services and administering sacraments except those for the sick and dying. The doge reacted by forbidding the promulgation of the interdict and instructing religious orders to ignore the pope's instructions or risk the loss of their property. All obeyed, with the exception of the Jesuits, who consequently were expelled from the city: their church and school were confiscated and given to nuns of the Benedictine order. The Society remained banned for over fifty years, and was able to return to Venice only in 1657 (Bousama 1968: 345–6).

Search for a Patron

Strachan remained only a week or two before moving on. En route to Paris, he called on Giovenale Ancina, the Bishop of Saluzzo in Piedmont, who wrote some flattering words in his album, describing Strachan as a 'Scot of outstanding erudition, pious and a friend' (Strachan's *Album Amicorum*: 24). The visit may have been merely a convenient stop on his journey, or a wish to renew a friendship which had been initiated two years earlier in Rome on the occasion of the bishop's ordination. It is likely also that Strachan was hoping for some preferment. Ancina was a distinguished classical scholar who, on taking up his appointment, embarked on a programme of major expansion and improvement of his diocesan seminary. His plan required more and better educated professors, and this may have given Strachan cause for hope. From his entry in the album it is clear Ancina had a high regard for his Scottish friend, but whatever the reason for Strachan's visit to Piedmont he did not take up a position there, and his stay was brief. For no discernible reason, he left Saluzzo to return the way he had come. Researchers into Strachan's life will constantly face what appear to be arbitrary actions of this kind on his part. Sometimes the most obvious conclusion to draw is that the young Scotsman enjoyed visiting new and interesting places. On his way to Paris, his diversion to Regensburg, the imperial free city in Bavaria, can be viewed in this light.

He arrived there some time before 28 March. It has been suggested that he had gone to visit the Scottish Benedictine monastery of St James in the hope of finding a position and that he 'taught to great applause' (Dellavida 1956: 13). This is unlikely to have been the case. He could not have taught at the monastery since the monks no longer ran the college for local youth, which the first abbot, Ninian Winzet, had set up in 1575. In 1590 the city authorities had invited Jesuits to open a gymnasium exclusively for the education of the youth of the city and, unable to compete with their superior level of teaching, the Scots Benedictines' establishment had closed. It appears that Strachan taught at the Jesuit gymnasium but he did call on his fellow countrymen in St James' monastery. Abbot James Whyte and Prior Adam Makcall wrote their names in the album amicorum. Strachan would have received hospitality from the monks, but this could not have been his reason for such a pronounced diversion from the road to Paris. Again, he may have been commissioned by Aquaviva to deliver a message to the Jesuits in the city. His stay in Regensburg was brief, and on his departure he went directly to Paris (Strachan's *Album Amicorum*: 13).

Despite any possible commissions from General Aquaviva, there are clear signs throughout his travels that Strachan was looking for gainful employment. On his arrival in Paris, an element of urgency appears to have entered his search. He immediately matriculated at the university, which was essential for any participation in academic life. In addition, he was required to join one of the four 'nations' which constituted the corpus of the University of Paris. He enrolled as a master in the German nation (the only one open to Scots). Its records show that for the next three years, he received payments as a preceptor specialising in teaching mathematics (Durkan 1971: 12). This involved taking occasional classes and tutoring wealthy students willing to pay for private tuition. Strachan may have gained the greater part of his income from tutoring individuals. In one entry in the album Jerôme Coignet, who describes himself as a member of a family of great political influence, calls Strachan his preceptor. Coignet used a whole page of the album to draw his elaborate family coat of arms and a stylised IHS symbol signifying some connection with the Society of Jesus. The entry is undated but its position on the page would indicate that it postdates 1606 (Strachan's *Album Amicorum*: 132). By itself this work could have provided him with only a modest income. According to Thomas Dempster, he also taught humanities at the Collège du Mans, which would have been equally poorly paid. The college was one of the least endowed in the university and closed eight years later due to lack of funds. Strachan was in need of a patron, and he continued to look for one in the Church.

Shortly after Strachan returned to Paris, Camillo Borghese was elected pope, taking the name Paul V. Strachan had known him in 1601 in Rome when he was the cardinal protector of Scotland. Making use of Borghese's elevation to the papacy, Strachan reintroduced himself and attempted to gain his patronage by composing and having printed a congratulatory poem for the occasion. The poem, 'Faustissimam Sanctissimi D(omini) N(ostri) Pauli V Pont. Max. inaugurationen, orbi christiano gratulatur Georgius Strachanus', was published on 16 May 1605 (Dellavida 1956: 18–19). In it, as well as delivering the customary flattery, he reminded Pope Paul of an earlier poem in his praise – as Cardinal Borghese – which the Scot had presented to him, and of the kindness which the cardinal had shown Strachan at the time. However, the second poem praising Borghese was also dedicated to the newly arrived papal nuncio in Paris, Maffeo Barberini, who was in 1623 to become Pope Urban VIII. Strachan was unknown to Barberini, and he used the presentation of the printed volume as an opportunity to introduce himself. The pope appears to have been impervious to the flattery, but Barberini responded positively. In 1606, when Barberini was raised to the rank of cardinal, Strachan dedicated his translation of Lucian's work on calumny to him (Strachan 1607) and addressed the new cardinal as '*patrono charissimo*' – dearest patron. In the book's dedication Strachan stressed the sincerity of his praise for the cardinal, a fact which could not be disputed because, as he wrote, this was in keeping with '*si consideres Scotorum integritatem*' (the truthfulness and honesty inherent in Scottish people) (BAV Ms. Barb. Lat. 2081). The presentation copy is expensively produced, bound in red leather and with gold tooling of the Barberini coat of arms on the front and back, and has a dedication date of 12 October 1606 (Dellavida 1956: 20).

Strachan obtained the nuncio's patronage and financial benefit, which, when added to his other earnings, allowed him to continue his humanist scholarship, writing poems and translating classical texts. The following years in Paris were to prove productive. The most significant surviving work was written in 1606 in memory of young Patrick Seton who had died the previous year, probably of the plague, at the Scots College in Rome. In this poem, 'Lacrymae' ('Tears'), he expresses himself movingly. It is clear that he was greatly affected by the death of his young friend, and the verses he produced are tender in character, much different in style from the stiff classical form that he normally used. The poem, which was published in Paris, was well received by the community of humanist scholars. 'Lacrymae' enhanced his reputation as a writer and scholar in the Republic of Letters.

Over the years that followed, Strachan enlarged his circle of influential acquaintances, becoming well known both in the city and at the royal court. In this he was helped by fellow Scots, who were well represented in the city. John Barclay, his friend from the time in Pont-à-Mousson, renewed Strachan's acquaintance. William Barclay (unrelated to John), a fellow Scots classicist, wrote two pieces of verse praising both 'Lacrymae' and Strachan (Leask 1910: 3, 9–10). Later, Thomas Dempster also expressed his admiration for the work, and it is him we have to thank for preserving a list of Strachan's most prominent works during his stay in Paris (Dempster 1627: 601). For the most part, these are translations into Latin of the works of ancient Greek writers – Agatharchides, Antiphon, Diogenes and Praxagoras are among those listed. Dempster also recorded that he had asked Strachan to translate work by another minor classical author, Vindanius Anatolius. Strachan had accessed most of these works from the *Bibliotheca* of the ninth-century patriarch of Byzantium, Photius; this work had hitherto been available to Western European scholars only in Greek. While Strachan was engaged in his translation, in 1606 the Flemish Jesuit, Andreas Schott, published a comprehensive Latin translation of *Bibliotheca.* He had been greatly helped in his work by drawing on an edited version of the Greek text by David Hoeschel, one of the prominent Augsburg humanists. Schott's book was a great success, and lessened the impact of later translations such as Strachan's. By working on these texts Strachan had kept himself at the forefront of scholarly research. What is clear is that little of the large volume of work that Dempster lists for Strachan was ever published, and the translations probably only served as new and interesting texts for his students at the Collège du Mans. Strachan also produced poetry, but what survives in print is either included in the work of others or attached to appeals for support to prominent churchmen.

In 1608 Cardinal Barberini was recalled from Paris having been made Bishop of Spoleto and cardinal protector of Scotland. Barberini's sponsorship of George Strachan did not end with his departure to Italy: indeed it appears to have been enhanced. Throughout his life Barberini was a great patron of the arts, and his support for Strachan can be seen in this light, but he also had use for the Scot in Paris. Letters from Strachan survive which show that he acted as Barberini's eyes and ears, and kept him informed of political developments especially those relating to James VI/I in Britain (BAV Ms. Barb. Lat. 2190). As cardinal protector, Barberini had a responsibility for the Catholics of Scotland. Following the Gunpowder Plot of 1605, matters had become particularly difficult for all British Catholics. There was an influx of exiles into Paris due to increased persecution at home, and Strachan appears to have tried to help some. One of their number was a young man, David Chambers, who later became the rector of the Scots College in Paris. Strachan befriended him, and in a letter dated 15 July 1609 he wrote to Barberini explaining the situation in Britain and commending the young Scotsman. This information was followed by a request for the cardinal to help Chambers, who was intent on visiting Rome (Dellavida 1956: 25).

Barbarini's continued support for Strachan allowed him to cease teaching at the Collège du Mans, but he still needed to tutor private clients (Durkan 1971: 12). His most notable student was John Casaubon, the eldest son of Isaac, one of the most prominent humanist scholars in Paris. The Casaubons were Calvinist, but in August 1610 John became a Catholic. Strachan was believed to have been instrumental in persuading him to convert, and Isaac was even accused by Herbert Rosweyde, a Dutch Jesuit, of deliberately arranging for Strachan to be John's tutor with the intention of instructing him in Catholicism under the guise of teaching mathematics. Casaubon had been under the protection of King Henri IV but, shortly after the king's assassination and his son's conversion, Casaubon accepted an invitation from King James VI/I to settle at his court in England where he died in 1614 (Dellavida 1956: 37).

Disillusionment

Contemporary accounts of his association with Casaubon erroneously describe Strachan as a Jesuit. Given his earlier association with the Society of Jesus and Cardinal Barberini's patronage, the mistake is understandable. However, the Scot's relationship with the hierarchy of the Church had been placed under strain. Despite Barberini's continuing financial support, Strachan had made a number of failed attempts to gain favour with the pope. In a letter to Barbarini dated 29 April 1609 he went so far as to criticise Pope Paul V (Dellavida 1956: 27). This was a remarkable change of view considering that he had earlier described Paul, when he was Cardinal Camillo Borghese, as his patron and lavished praise on him in a poem on his elevation to the papacy. The reason for the change in attitude was that Paul had ignored his requests for a paid appointment as a papal agent in Paris. He accused the pope of being indifferent to Scotland and the welfare of the Scottish exiles in Paris – both Catholic and heretic. In his letter, he said that the need for support of the former was obvious but he argued that the latter were equally in need of instruction in the true faith, a task that he offered to undertake. This comment would appear to give credence to the rumour that he had been instrumental in the conversion of John Casaubon. The letter also makes clear that he had appealed previously on several occasions, to no avail.

His purpose in writing to Barberini was to ask him to intercede with the pope on his behalf. The Scottish agent in Paris, Mr Fraser, who was tasked with looking after Scottish exiles, had just died. Strachan wanted to replace him or, if the pope preferred, to act as papal envoy to King James in England. He suggested that, as a fellow Scot, he would be able to progress the Catholic cause with James better than the Jesuits, who had done so much to alienate the king from Rome. He may also have thought himself well suited due to his mission to James on behalf of Bishop Chisholm seven years before, but he made no mention of the fact that King James had exiled him shortly afterwards. His criticism of the Jesuits indicates that his attitude had changed fundamentally from the period when he had considered joining the Society. His earlier involvement with the Jesuits had ended, and by 1609 in Paris he was no longer acting for them in any capacity. Strachan had started his letter to the cardinal with an apology for the long delay in writing which had been caused by illness. The whole mood of his communication was one of dejection. It contrasted markedly with the letter of the previous year, when he was commending Chambers to his patron. The impression it must have created on Barberini was negative and it is unlikely to have been received favourably. It is not known if the cardinal did speak to Pope Paul on Strachan's behalf but any effort on his part could only have been half-hearted at best. The following year Strachan wrote to him again, stating in a letter dated 10 June 1610 that he had received a refusal of the post of Scottish agent in Paris from the papal secretary, Cardinal Datary, on the grounds that no money was available (Dellavida 1956: 34–6).

His growing disillusionment with the Church caused Strachan to increase his efforts to find a patron elsewhere. Through his involvement with the scholarly community in Paris and his connections to the papal nuncio, he had gained entry to King Henri IV's court and he tried to capitalise on this. It is known from an account given by Cardinal Jacques Dary du Perron that Strachan was held in high regard by the king and that he came close to obtaining royal patronage. The cardinal wrote in his diary of a conversation that he had had with the king:

> Il me dit un jour parlant de Monsieur Strachan, que c'est le plus honnête Ecossois qu'il ait jamais vû, et qu'il falloit luy faire avoir une place entre ceux qui doivent discourir devant le Roy. (He told me one day speaking of Mr Strachan that he is the most honest Scot whom he had ever seen and that a place must be found for him among those who are expected to make speeches in the presence of the king.) (Dellavida 1956: 36)

With such a recommendation, Strachan undoubtedly would have received a position at court but, unfortunately for the Scot and even more so for the king, Henri was assassinated in May 1610. According to Dempster, Strachan then abandoned the royal court but later obtained a position in the household of the king's cousin, Charles of Lorraine, 4th Duke of Guise, as the duke's mathematician. By this time he had lost the patronage of Cardinal Barberini. His final letter of complaint appears to have ended the relationship between the two men. The duke's patronage was an inadequate replacement for the cardinal's generosity and Strachan was not happy with his new arrangement. He restarted work at the University of Paris as a preceptor to supplement his income. It appears that he was helped in this by Cardinal Perron, the same courtier in whom King Henri had confided his respect for Strachan (Dellavida 1956: 38).

His friend Thomas Dempster, who had returned to Paris in 1609, was in a similar situation teaching at the university and looking for a sponsor. In 1613 he published a revised edition of Johannes Rosinus's *Antiquitatum Romanorum Corpus Absolutissimum* and dedicated it to King James VI/I in England. Strachan provided an epigram for the book, praising the scholarship of his friend. Dempster, in turn, included Strachan in the dedication to the king, describing him and other Scots Catholic exiles as:

> 'alumni of the Muses', guests of France and professors of French universities through the kind solicitude of Cardinal du Perron, who would be happy to serve the King of England and Scotland, if he would recall them to their country.

The book impressed James and he invited Dempster to his court, giving him the position of Historiographer Royal; Dempster was the first person to carry that title in Britain. There was, however, no invitation for Strachan who remained in France in the service of the Duke of Guise.

A Chance Encounter

If Strachan had carried out a review of his career at this point in his life, it would have caused him pain. He was over forty years of age and could no longer consider himself a young man. It had been seven years since he had received universal praise for his poem 'Lacrymae', which was to be his greatest literary success and, despite the fact that he was extremely erudite and generally found it easy to make friends, his career was in the doldrums. The allowance he was receiving from his sponsor, the Duke of Guise, was insufficient to provide for his needs. He had been reduced to returning to class teaching in Paris to supplement his finances. We know from Thomas Dempster's account that he was unhappy with his life in the duke's household (Dempster 1627: 601). Any chance of gaining a rewarding commission from the Church had disappeared and, following Dempster's failed attempt to influence King James on his behalf, his sentence of exile from

Scotland remained firmly in place. Younger friends such as Dempster and Barclay were being rewarded for their work in ways that appeared to be closed to him. He could have seen no prospect of a transformation in his fortunes. At this low point he decided that a complete change in his way of life was necessary.

The decision appears to have been triggered by a visit he received from a fellow countryman. In the summer of 1613, the household of the Duke of Guise including Strachan was resident in Aix-en-Provence. It was normal for the nobility to leave Paris in summer to avoid the heat and pestilence of the city. As he was its provincial governor, the duke chose the capital of Provence as his summer residence. William Lithgow of Lanark stopped in Aix while travelling from Paris to Italy. Lithgow was ten years Strachan's junior, and had recently returned from a three-year journey around the Middle East, the first of three that he was to make. During his stay in Provence, he found Strachan to be convivial company and spent time with him discussing his adventures. From Lithgow's account of his life and travels (Lithgow 1632) we know that Strachan showed great interest in what he had to say. Among other things, the men talked of Lithgow's visit to the ruins of Troy. Strachan's imagination was fired when the traveller gave him one of three ancient coins which he had found while walking over the site of classical Ilium (Bosworth 2007: 50). It took little more than this encounter with Lithgow for Strachan to decide that he wanted to travel through the Middle East. He immediately resigned his position as mathematician to the duke, returned to Paris to collect the last payment due to him from the university and wrote to his young friend, Thomas Dempster, informing him of his decision to go east to learn oriental languages. Dempster was in England at James' court and Strachan may have included a note to be forwarded to his family in Scotland. If so, it is likely to have been Strachan's last letter to them. His chance encounter with Lithgow when he was at a low point in his fortunes was enough to send him on a path which was to establish him as a major scholar of the seventeenth century.

CHAPTER FOUR

To Constantinople

East–West Relations

In the early seventeenth century, when George Strachan made his decision to go to the Middle East, journeys by Europeans in the region, although limited to a few categories of travellers, were not unusual. From ancient times there had been interchange between Europe and the East. Obstacles to free movement had appeared with the rise of Islam, and the nature of the relationship between East and West changed again with the Crusades. Following the final extinction of the Crusader kingdoms which occurred with the fall of Acre in 1291, the way in which the two cultures of Islam and Christianity viewed each other went through more transformations. The continuously changing nature of the relationship between East and West has been the subject of much debate among scholars, and different theories regarding the interactions have been developed. These include that of the 'Global Village' proffered by Fernand Braudel (Braudel 1972–3), and an almost diametrically opposite view of a mutual exclusion between the Islamic and the Christian worlds that has been described by James G. Harper as the 'Iron Curtain' model. Edward Said added to the debate by introducing the concept of Orientalism to identify the changing imbalance of power which defined the relationship:

> The orient is not only adjacent to Europe, it is also the place of Europe's greatest and richest and oldest colonies, the source of its civilisations and languages, its cultural contestants, and one of its deepest and most recurring images of the Other. (Said 1978: I)

Harper has recently refocused attention on the problem by stating that neither the 'Iron Curtain' model nor the 'Global Village' theory adequately explains these relationships. East and West were never entirely closed or entirely open to one another in all respects but altered their mutual responses as circumstances dictated (Harper 2011: 5–6).

No matter the view taken, it is generally accepted that in the Early Modern Period, when Strachan went east, free movement was possible. Visitors from the West continued to go east both as pilgrims to the Christian sites of the Holy Land and as merchants trading with the Levant and beyond. Also, numerous as these visitors were, possibly the greatest European presence in the East from the late Middle Ages onwards was that of those who had been taken as slaves. At that time the use of slaves in Europe had declined, but in the Middle East slavery continued to thrive. Islam forbade its adherents from enslaving fellow Muslims, although, despite its prohibition in the Qur'an, Sunni Muslims enslaved Shi'a Muslims whom they considered heretical (Zarinebaf 2011: 28). The demand for slaves was satisfied mainly using African Animists and European Christians. Many were captured in raids by pirates or were the booty gained in wars of conquest. Arab and Turkish dealers predominated, although Europeans also were engaged in trading slaves. Most countries around the Mediterranean Sea participated to some extent, but the Genoese in particular were heavily involved, having monopolised the trade at Caffa on the shores of the Black Sea where Mongol, Tatar, Russian and Circassian slaves were sold. When Turkish forces drove them out of the port in the middle of the fifteenth century, Italian merchants continued to buy these slaves and transport them to Egypt, where Circassians in particular were much prized (Wolff 2003: 3). Meeting the Eastern world's demand for slaves became one of the drivers of the West's trade with the East.

Christian Europe's interest in the East was centred on trading in luxury commodities. Throughout Europe a significant market existed among the elite for spices, fine silks and cottons which came from India, and the Middle and Far East. Until the Portuguese discovered the sea route around the Cape of Good Hope to India and the Spice Islands, all of this merchandise had to pass through Muslim lands to reach the markets of Europe. Even after the route around the Cape was opened up, for many years most trading was still conducted overland to the Mediterranean Sea, as it had been since ancient times. The East had long lost its monopoly on silk production which is believed to have originated in China. Knowledge of its technology had spread across much of the Old World with production being carried on in places where it was possible to grow the mulberry trees on which the silk worms fed. Sericulture had been introduced to the Byzantine Empire before the reign of Justinian I in the sixth century and later spread throughout the Muslim world (Jacoby 2004: 198). By the end of the thirteenth century, the cities of northern Italy – Lucca, Venice, Genoa, Pisa and especially Florence – had grown wealthy on their skill in the production of fine silks and brocades (Mainoni 2000: 365–98). Silk production was also important in Spain with the Morisco community continuing their long-established manufactures even after the Reconquista (*El País* 2013). When the Moriscos were expelled in 1571, silk production in the south of Spain ended, but by then the city of Valencia had become a significant centre for silk manufacture due to Genoese artisans being induced with financial incentives to settle there (Marco 2018: 117–18). In northern Europe the Flemish had the most successful silk manufactures. They achieved this by importing raw silk and processing it into rich brocades. These, along with fine woollen textiles produced in Flanders, were often re-exported to the East where Mamluk Egypt, in particular, became an important market (Wolff 2003: 7). Despite the high quality which European industries produced, fine silks in the form of fabric and carpets were still imported from the East with China and India being dominant suppliers to the Muslim world and European market.

Venetian Dominance

Although the demand for Eastern merchandise was Europe-wide, not every country was in a position to trade directly with the Islamic world and, until the sea route around southern Africa was opened up, none could trade directly with India and China. Before the Ottoman Turks defeated the Mamluks and brought Egypt, Palestine and Syria into their empire in 1517, the Republic of Venice had become the main conduit to and from the East, and was the dominant Western trading partner of the Ottoman Empire. From small beginnings in late antiquity, the Venetians had accrued advantages in this trade. These were greatly enhanced with the sacking of Constantinople in 1203 during the Fourth Crusade. The Byzantine territories, which they gained from the emperor, gave them strategic ports and islands in the Aegean that allowed them to secure their trade routes to the East. These trading arrangements remained in place even when the Ottomans began to contest these Venetian colonies in the sixteenth century, weakening Venice's position militarily. The Ottoman capture of Cyprus in 1571 demonstrates this point. The island's political suzerainty changed, but the leading Venetian families remained on Cyprus under Ottoman rule, plying the same trade with the same business partners, Muslim and Christian (Rothman 2012: 59). The island's loss did not deprive the Venetians of their staging post to the Levant, and Venetian connections to the Middle East remained the most extensive and robust among the European powers.

The ongoing debate among academics regarding the nature of East-West relationships is limited in the amount of light it can throw on the experiences of George Strachan. He did not belong to any of the categories of traders, pilgrims or slaves that constituted the greatest European presence in the East. He was one of a small number who went east for the purpose of self-improvement. This group was made up almost entirely of educated and wealthy men. In earlier times they would have gone on religious pilgrimage but, in the turmoil of the Reformation and Counter-Reformation, this practice had come under criticism in both Catholic and Protestant Europe. The perceived advantages of foreign travel were still desired – the acquisition of political and social experiences, the perfection of languages and a level of sophistication that could not be obtained by staying at home (Carey 2009: 2). To obtain these benefits, some members of social elites embarked upon secular journeys and the truly adventurous did not restrict their travel within Europe. Strachan can be viewed as one of this select group, but the Scotsman stands out for a number of reasons. He was older than most adventurers and his earlier travels around Europe had already given him a cosmopolitan polish. His reasons for travel were principally scholarly and he remained much longer and visited many more regions than others. As a traveller in the East, Strachan was unique.

There is no hard evidence surviving of how he travelled to the East. He left no record of the route he chose but, like other travellers, he would have known that his most practical means of getting there was on a Venetian ship. It is known, from Lithgow's account of their meeting in Aix-en-Provence and Dempster's short biography of Strachan, that in the summer of 1613 while in the south of France he made up his mind to go to the Orient to learn Eastern languages. From what followed it is clear that he moved quickly once he had taken his decision. First he went to Paris and collected his unpaid earnings from the university (Durkan 1971: 12). While there it is likely that he lodged at the Scots College. He left his album amicorum and possibly other belongings that he did not want to take on the journey in the keeping of its rector, William Lumsden (Halloran 2003: 28). He had stopped using his album in 1609; the last entry was made by Peter Hay on 5 September of that year (McRoberts 1952: 116). He did not stay long, since he needed to reach Venice before the end of autumn when the regular sailing season to the East would be over for the year.

Practical Considerations

Before attempting to follow Strachan's progress any further, it is worth considering how the relatively impecunious Scot was intending to pay for what was going to be an expensive expedition. He was about forty years old and had spent the greater part of his adult life earning his living in Europe by teaching and writing. He had acquired a first-class education, and had maintained himself as a gentleman in line with his upbringing as a member of the Scottish nobility. Nevertheless, he had not accumulated any wealth and, as a consequence, had to acquire the knowledge of how to live inexpensively. On occasion he must have made use of *mensam ambulatorium* (the walker's table), which involved a guest timing his visits to mealtimes. It is a technique by no means unknown today but, in Strachan's time, Scottish manners required that hosts, no matter how straitened their circumstances, received all such visitors. In turn guests were required to be profuse in their gratitude and praise their hosts. The accounts given of him in the entries in his album amicorum show that Strachan had an amiable disposition, and it is likely that he was seen by his hosts as an entertaining and welcome guest. Earlier in his career, as he travelled through France, Spain, Italy, Germany, Switzerland and the Low Countries, he had found accommodation in colleges and religious foundations. His stays in the Pontifical Scots College in Rome and the Schottenkloster in Regensburg show that his status as an eminent scholar allowed him free board in most if not all such establishments. Normally visitors were not charged, but would give a donation according to their means. As a near penniless scholar, Strachan would most likely have thanked his hosts by leaving the gift of a small poem of praise rather than money. Religious and educational institutions which operated on this basis were available to travellers throughout the Continent, especially in Catholic Europe. On his journey to the Middle East, Strachan could not rely on the same benign arrangements everywhere but, although the circumstances were different, a number of religious orders maintained a presence in the Muslim lands which supported Christian pilgrims. Strachan would have known of these and could count on their hospitality. In addition, during his career Strachan had cultivated a large number of influential friends and acquaintances from whom he obtained hospitality. These had proved to have been of considerable help to him while he lived in Europe, but he could not have expected their help to continue during his travels in the East. Surprisingly, as will be seen, he was able to make use of them even there.

During his conversations with William Lithgow, the subject of how he had afforded to travel extensively is likely to have arisen. Lithgow was not a wealthy man, but had managed to support himself for three years while journeying in the East. From their discussions Strachan knew of the book that Lithgow was about to publish giving an account of his travels. The gentleman from Lanark hoped that it would fund the future journeys he was already planning. Given that Strachan was by that time a published author, it would have been reasonable for him to think that such a commercial opportunity would be available to him once he returned from his pilgrimage. The appetite among Western scholars for books on the East had existed for centuries. Cyriaco de' Pizzicoli, a merchant of Ancona, had written as early as the twelfth century, at the request of Pope Eugenius IV, telling of his travels in Egypt (Wolff 2003: 64). Many others wrote of pilgrimages to the holy sites of Jerusalem (Santo Brasca 1481). In addition, books of a more general nature describing the East were popular (Nicolay 1576). As a well-read scholar, Strachan would have been aware of a number of these published works. When these points are taken into account, the fact still remains that Strachan was taking an enormous risk that he would run out of money on his journey and be stranded in the East. The driving force of dissatisfaction with his current life must have been great. Together with his natural curiosity, it was enough to launch him on his great adventure.

Pietro Della Valle

All that is known for certain about Strachan for the first two years after he left Paris is that he went to the Levant, that he was in Constantinople in 1614 and that he had also visited Mount Lebanon, the source of the cedars of Lebanon mentioned in the Bible. These details are found in one of the major sources of information on Strachan's time in the East (Della Valle 1664: 718). This limited information was given by an Italian traveller who befriended the Scot when they met in Persia. Pietro Della Valle was one of the small group of educated and rich travellers who had gone east for self-improvement. He was a Roman nobleman who had been thwarted in love, and reacted by indulging himself in a trip to the Orient which started in 1614 with the purpose of seeing as many interesting places as possible. Della Valle considered his journey to be a pilgrimage, but his behaviour was more akin to that of someone conducting a Grand Tour of the East. Later, it became fashionable for young noblemen to view the classical ruins of Rome and Greece, and his journey can be seen as a forerunner of this new diversion for young men from wealthy families. He spent twelve years travelling around the Ottoman, Safavid and Mogul Empires before returning to Italy in 1626. During this time, he wrote an extensive diary in the form of letters which he sent home to a friend, Doctor Mario Schiapano, a university professor. His writings follow the advice given in Albrecht Meir's *Methodus describendi regions, urbes et arces*, published in 1587. Meir wrote that the traveller should ensure his observations covered cosmography, geography, topography, husbandry, navigation, the political and ecclesiastical state, and also 'literature' (Rassen 1994: 160–86). Doctor Schiapano had agreed to edit the letters with the intention that, on Della Valle's return, they would publish the account of his travels (Bull 1989: xvii). Schiapano did not complete the editing, and the only account published in Della Valle's lifetime was that of his travels in Turkey. The doctor's failure in this regard is understandable, since the letters contain over three million words. A fuller edition of the letters was published by Della Valle's sons in 1664. Translations in a number of European languages, of the fuller accounts, have been published as scholars have come to appreciate the importance of the wealth of information on the East contained in the Italian's account.

Prior to meeting Strachan in Persia, Della Valle had begun his voyage by sailing from Venice to Constantinople and from there to Alexandria in Egypt. He had travelled from Venice with an entourage which consisted of two servants, a Flemish painter, whose task was to produce illustrations for the intended book, and an Augustinian friar who was to be his personal confessor. In Constantinople he added to these with an additional painter, an interpreter and an Ottoman official to act as a government minder. The official had his own servants in attendance, and accompanied him throughout his stay in Ottoman territory. Whether the official's duty was to ensure that Della Valle encountered no problems or that he did not create any was not clear even to Della Valle. This enlarged party travelled up the Nile to Cairo to visit the pyramids, where Della Valle carved his name on the pyramid of Cheops before conducting some private archaeological excavations. He bought, as souvenirs, two human mummies which are now in the collection of the Dresden Art Gallery (Bull 1989: xii) and departed Egypt by joining the annual Hajj pilgrimage to Mecca. His party left the pilgrim caravan at St Catherine's monastery in the Sinai Desert and travelled through Jordan to the Holy Land. From there he went to Aleppo with the intention of travelling on to Persia. It was after leaving Aleppo that he first heard reports of 'Signore Strachan', the Scottish gentleman in the desert, and decided he would like to meet him. This did not happen on that occasion, however, and Della Valle moved on to Baghdad where he met and married a young woman of Christian-Arab and Armenian descent. After an understandable delay in Baghdad, he and his new bride, together with his remaining servants and many of his wife's extended family, travelled on to Isfahan, the new Persian capital. It was there that he met Strachan. The two formed a firm friendship which lasted for more than two years while Della Valle remained in Persia. This brief account of Della Valle's journey is given in contrast with what we know of Strachan's journey. In the Roman's case, he left a prodigious volume of writings, and there was never any lack of money to cover his many expenses. On the other hand, for the Scotsman, as will be seen, money was an ever-present concern, and he left very little in writing concerning his travels.

Travelling to the Ottoman Empire

Strachan's first destination on leaving Paris in the summer of 1613 was almost certainly Venice. On his arrival, he did not have the possibility of lodging at the Jesuit school where he had stayed on his previous visit in 1606. It had become a convent for Benedictine nuns. Nevertheless, he was familiar with the city, and would have found accommodation readily available at one of a number of monasteries. His task was to find a suitable ship to take him east and, like any traveller, he would have had a number of options depending on his purse and choice of destination. The most important trade between La Serenissima and the Ottoman Empire was with the city of Alexandria in Egypt. The Venetian senate organised two large argosies sailing to the Egyptian port each year. Because their cargoes were extremely valuable, the fleets were accompanied by a number of armed warships. The majority of passengers were merchants trading between Europe and Alexandria. Occasionally a few rich travellers, like Della Valle, who were visiting the sights of Egypt's ancient civilisation, joined these argosies. The added security provided by the armed convoys came at a high price, which only the wealthy could afford.

The *signoria* also organised other sailings eastwards from Venice that ran on a twice-monthly basis. These were part of the republic's postal service to Constantinople and other major cities of the empire, especially Aleppo. The postal ships were fast galleys which sailed to the small Venetian port of Cattaro (Kotor in present-day Montenegro), a distance of 500 miles which the ships could cover in an average of two weeks. From there the mail was taken overland by Montenegrin couriers who, although they were paid employees of the Venetians, were subjects of the Ottoman sultan, a status which allowed them to travel without the need for special permits. Their route took them through dangerous territory, and the couriers travelled in armed bands. This road was used in preference to the onward sea route to

Constantinople because delivery times were faster (normally fifteen days to cover a distance of 650 miles), and the service could be relied upon in a way that was not usual with sea voyages to Constantinople. By confining the postal galleys to the Adriatic Sea, winter sailings were rarely interrupted by bad weather unlike in the more open waters of the Eastern Mediterranean (Dursteler 2009: 601–23). It would have been rare for travellers from Venice to have accompanied the post unless their ultimate destination was Cattaro. Although Strachan did not use this route, it appears that he was aware of it and he later used the service to send letters back to friends in Europe (Dempster 1829: vol. 2, 601).

The realistic option for travel to the East which was readily open to travellers was, as taken by Pietro Della Valle and his party, to sail on whatever ship was leaving for their intended destination in Egypt, the Levant or to Constantinople. It is known from Della Valle's journal that on leaving Europe, Strachan's first went to the Levant (Della Valle 1664: vol. 2, 437). In Della Valle's case, he had determined to go to the capital of the Ottoman Empire itself. Accordingly, he boarded *Il Gran Delfino*, a Venetian war galleon armed with forty-five cannon, which made a number of port calls on the way to Constantinople. The galleon was acting as a passenger ship, having over 500 people on board. As well as its crew of sailors and soldiers there were merchants and general travellers of both sexes. Della Valle categorised them as 'Catholic Christians, heretics of various sects, Greeks, Armenians, Turks, Persians, Jews, Italians from almost all cities, French, Spanish, Portuguese, English, Germans, Flemings and to conclude in a few words, [people] of almost all religions, and nations of the world' (Della Valle 1664: vol. 1, 1–2; Dursteler 2006: 1–2).

Della Valle had chosen Constantinople as his initial destination because he wanted to obtain a permit to travel widely in the Ottoman lands. Permits issued by the sultan's court were recognised throughout the empire. Otherwise, the Roman would have required a new permit for each of the provinces he visited. As it turned out, the permit came with the accompanying Ottoman official to supervise his actions.

George Strachan was in a different position. Given that his first destination was the Levant, he would have sailed on one of the many ships that took Christian pilgrims to the Holy Land. During the pilgrimage season, these sailings were frequent and represented the least expensive means of travel to the Orient. Pilgrims needed only a limited permit that they could obtain on arrival at their first port of call, which was normally Acre. The city was a few miles from Jerusalem and was the principal pilgrimage port. Della Valle gives no details, but it can reasonably be assumed that Strachan took this route, since it is inconceivable that he would not have included in his plans a visit to the holy sites of Christianity, especially Jerusalem.

Dempster wrote that Strachan spent six years travelling through the Holy Land, but in this he is mistaken since Strachan was in Aleppo in 1615. This was less than two years after he left France, having visited Constantinople and Mount Lebanon on the way (Dempster 1829: vol. 2, 601) (Chapter 5). There were frequent sailings of merchant ships to the Levant ports, and Strachan would not have waited long in Venice before obtaining passage. By the autumn of 1613, he was in the East and most likely in Jerusalem.

The Holy City

The Jerusalem he found has been described by George Sandys, who visited the city in 1610. Sandys was the son of the Anglican Archbishop of York and can be viewed as another of the group of educated young men who travelled east for self-improvement. As a Protestant, before setting off on his journey he would have received warnings regarding the dangers to his soul that he faced. Exposure to the corruption he would meet travelling south and east 'into the jawes of danger: for so farr hath Satans policy prevailed' were to be guarded against. Some commentators advised against travel outside Protestant lands, although others wrote that anyone who had 'sucked the pure milke of true Religion, and Orthodoxall truth' would never be corrupted (Carey 2009: 9–11). Sandys was on his guard and his writings portray a jaundiced picture of the other religions he encountered.

The city he visited had endured centuries of siege and warfare following Saladin's defeat of the Crusader kingdom of Jerusalem in 1187 and its buildings had suffered accordingly. Sandy's description was of a city 'much of which lies in waste, old buildings all ruined, the new, contemptible' (Sandys 1621: 157–8). The new constructions that Sandys found contemptible were the work of Sultan Suleiman the Magnificent carried out in the middle of the sixteenth century. As keeper of the holy sites of Islam, he saw it as his duty to restore Jerusalem as a major Islamic city. Apart from rebuilding the city walls and renovating the Dome of the Rock, the sultan installed new gateways and added several new mosques. One of his major improvements was the construction of an aqueduct with nine fountains to provide drinking water in every district in the city. By the end of Suleiman's reign in 1566, the city's population had trebled to 16,000, the great majority of whom were Muslim (Montefiore 2011: 292–3). Sandys' contempt for this work was based largely on his attitude to Islam rather than on the quality of the buildings many of which still grace the old part of today's Jerusalem (Haynes 1986). George Strachan is unlikely to have viewed the city in the same light.

Jerusalem is regarded as holy by three faiths, but Suleiman's major reconstruction work was part of his plan to make Islam the dominant religion in the city. In pursuit of this objective, he acted to cause dissension among non-Muslims by giving the French Franciscans a privileged position among the many Christian denominations. He declared the Franciscans to be the principal custodians of the Christian shrines and allowed them to set up their headquarters in the convent of St Saviour (known in Arabic as 'Dayr al-Latin'), which was situated close to the holiest Christian site of the

Church of the Holy Sepulchre (Montefiore 2011: 296). Originally founded in 1230, the Franciscan convent in Jerusalem is the centre of the Franciscan province of Sanctae Terrae (Holy Land, also called 'Siriae', or Promised Land) which covers all of the Levant and Syria as well as the island of Cyprus (Golubovich 1906: 356). The Franciscans turned their convent into a centre for Catholic pilgrims, and it is to this refuge that Strachan would have been required to turn on his arrival in Jerusalem.

In his account of his visit, George Sandys relates that he was met by two Franciscans outside the walls of Jerusalem and accompanied to their convent. They informed him that the Muslim authorities considered any Westerners who did not reside with them as potential spies (Sandys 1621: 154). The Protestant traveller paid the friars 100 dollars for his accommodation over four days which he thought was expensive. The Franciscans explained that such charges were necessary because the Turks took every opportunity to extract money from the Christians in the city, and that their convent was not exempt. In their writings Sandys and Della Valle used the term 'Turk' to describe all Muslims and in so doing they were following the European usage. In turn, Muslim writers referred to all non-Muslims as *kafirs* with Christians from the West being further described as Franks. The Franciscans told Sandys of a recent incident where a Turk had made a demand of one of the Franciscans which the friar had refused. The Turk then punched himself on the nose causing it to bleed, and immediately complained to the city authorities that the friar had assaulted him. There was no question of the court disputing the word of a Muslim against that of an infidel *kafir* and the Franciscans were fined 800 dollars (Sandys 1621: 159). Under such circumstances, it is unlikely that when Strachan stayed at the convent the Franciscans would have waived the cost of his accommodation, although the charge was perhaps less than that faced by their wealthy Protestant guest.

Pietro Della Valle had a similar experience when he arrived in Jerusalem in March 1616. He was met at the Jaffa Gate by the father vicar and a number of his fellow friars and, before passing into the city, was required to pay a toll to the Turkish officials which he said was greater for Christians than for others. On arrival at the convent, he was met by other officials who searched his baggage for dutiable items. Charges were levied and bribes exchanged hands. Della Valle relates that, because of the size of his entourage, the rumour had spread that he was the son of a king. As a result, he was visited by the city governor who tried to confiscate more of his property. The Ottoman functionary was dissuaded from doing so by the production of Della Valle's permit issued in Constantinople, and the presence of the sultan's official minder (Della Valle 1664: 423–5). While he was their guest, the Franciscans showed Della Valle all the Christian sites, the principal of which were the Way of the Cross and the Church of the Holy Sepulchre. When George Strachan visited, it is likely that he would have been afforded the same courtesy (Della Valle 1664: vol. 1, 428–9).

Sultan Suleiman's initial reason for nominating the Franciscans as guardians of the Christian sites was to help gain the alliance of the French king in his wars against Emperor Charles V; causing division among the Christian denominations was an additional benefit as it helped maintain Muslim dominance of the city. The representatives of the different Christian rites resented the Franciscans' position and vied with them for control, especially of the main shrine of the Holy Sepulchre. Armenians, Copts, Maronites, Nestorians, Jacobites, Abyssinians and Georgians all had chapels in the church (Sandys 1621: 160–70). However, the Greek Orthodox Church was the principal contender and claimed the *praedominium* (guardianship) for itself. The Orthodox clergy refused to recognise the Franciscans' authority, and asserted that the *praedominium* was not within the gift of the sultan but was theirs by right of long standing. Their resentment over this matter only added to their hatred of the Latin Church which stemmed from the Great Schism of the eleventh century and the Sack of Constantinople by the Crusader army in the thirteenth century. On the great feast days of the Church, it was not unknown for the representatives of the Latin and Eastern Churches to engage in fights in the basilica to determine which of them would have first use of the sanctuary to conduct their service. In the 1980s, more than three-and-a-half centuries later, the writer William Dalrymple found that the situation was largely unchanged. On being shown around the Church of the Holy Sepulchre by his guide, the Franciscan Brother Fabian, he was told that they could not enter part of the church until the Greek Orthodox monks left, which they would not do until the end of their allotted session. Asked if the different denominations cooperated with each other, Brother Fabian replied, 'I think the Copts are speaking to the Armenians but apart from that, no' (Dalrymple 1990: 4).

Into this confusing mix has to be added the Muslim family of the Nusseibeh. Following the Arabic conquest of Jerusalem in the seventh century, a family of that name was appointed custodians of the Temple Mount and the Church of the Holy Sepulchre, a position which they hold to the present day. They keep the keys to the church and control access to it. By charging entrance fees, they earn substantial amounts from visitors. In the early seventeenth century, at the time of Strachan's visit, Christian pilgrims represented the greatest source of income to the city. At Easter when visitor numbers were highest, the city governor came with the city's armed garrison to enforce his authority and charged an entrance fee of ten gold pieces to each of the 20,000 'hell-destined infidels' (Montefiore 2011: 298). To the Protestant visitor, George Sandys, such commercial behaviour was no better than that of a fairground huckster and was encouraged by the Orthodox and Catholics through their rituals; they had turned the city 'once glorious, elected by God for his seat [into a] theatre of mysteries and miracles' (Sandys 1621: 171).

Before Strachan set out on his journey eastwards, his knowledge of Semitic languages was limited to Hebrew and Chaldean. Hebrew, along with Latin and Greek, was required by scholars of the Bible. Known as *Sacrae Linguae*, the teaching of the three languages was an integral part of the *Ratio Studiorum*, the curriculum followed by all Jesuit colleges. Where possible, the Jesuit teachers were proficient in Chaldean. Strachan had mastered this tongue in addition to the other three (Chick 1939: 236). There is no record of his having knowledge of any other languages of the Churches of the Eastern Rites. Paradoxically, in Jerusalem his knowledge of Hebrew was useless among the city's Jews. Although there was a thriving community of about 2,000, they were Sephardic, descendants of those forced to leave Spain at the end of the fifteenth century, and spoke Ladino, a language with Arabic roots different from but arrived at in a similar way to the Germanic Yiddish language of the Ashkenazi Jews of Eastern Europe (Montefiore 2011: 293). Neither language corresponded to Hebrew in any mutually comprehensible way. Strachan would have been more familiar with their ancestral tongue than all but a handful of Jewish scholars in the city. Jerusalem's Jewish community stood apart from the Muslim and Christian populations. In his journal, George Sandys wrote that he viewed them with amusement and repulsion. Despite Jews being tolerated in the rest of the empire, in Jerusalem, as part of his policy of restraint on all religions other than Islam, Sultan Suleiman limited their public worship to the street in front of a short nine-foot section of the surviving wall of Herod's temple (Montefiore 2011: 298). They had built four synagogues in the quarter of the city allocated to them to accommodate their services, but the 'Wailing Wall' was the only part of the city where they could be observed worshipping by non-Jews. The wall was adjacent to the Jewish Quarter and when Sandys saw them praying there he thought it 'impossible not to laugh [because] their fantastical gestures exceed all barbarity with ridiculous nodding' (Sandys 1621: 157). Sandys' reaction would have been caused in part by the fact that he had not encountered any Jews prior to leaving England on his travels. Strachan is unlikely to have reacted in the same way, given that he had spent his adult life travelling through Europe and would have encountered a number of Jewish communities such as that in Venice. His experiences in Europe must have helped prepare him for his visit to the Middle East, where he was subjected to many new, strange sights. On his travels he showed by his behaviour that he was open-minded to the people he encountered and, as will be seen, actively embraced new cultures.

Visiting Classical Sites

In his journal, Della Valle makes no mention of Strachan's travels in the Levant, other than that he went there first. His silence on this may be no more than a reluctance to repeat descriptions of places he himself had visited. Despite this lack of detail, it is impossible to believe that Strachan would have stayed long in the febrile atmosphere of Jerusalem. He had come to learn Eastern languages and the city afforded few opportunities. The

Muslim community was predatory on Christian visitors and the Orthodox and Eastern Rites Christians were hostile to them. He would have had little chance of conversation with anyone other than the Franciscans and their European guests. In addition, his precious supply of money would not have allowed a prolonged stay. Since we know from Della Valle's account that he was in Constantinople in the spring of 1614, he must have left Jerusalem in late 1613 or early 1614.

Furthermore, it is clear that his intentions in travelling were not restricted to learning Eastern languages. As a young man, he had travelled through much of Europe, taking pleasure in visiting places of interest. No doubt this was driven by his natural curiosity, but such behaviour in the East would have fitted with his plan to write a travelogue on his return to Europe. After leaving Jerusalem, his journey to Constantinople must have been by ship. A journey overland would have proved too difficult for a lone traveller. It is unlikely that he would have been able to take a direct sailing from a port in the Levant to Constantinople. Such sailings were infrequent in winter. Due to the lateness of the season, Strachan's sea voyage is likely to have been by *saettia.* These small, fast pinnaces were three-masted, lateen-rigged sailing ships which were lightly armed and relied on their speed for safety, and were used ubiquitously in the Eastern Mediterranean for short journeys in inshore waters (Moshenka 2016: 220; Bruce 1868: 19). They had covered decks which allowed them to sail in much of the rough weather that winter brought. They remained in their home waters where the crew was familiar with the sailing hazards. In sailing by *saettia,* Strachan would have been required to change ship on a number of occasions when the captain reached the port at the limit of his sailing range. This means of travel gave Strachan the opportunity of visiting a number of places along the coast of Asia Minor. He would have been familiar with a number of these from his studies of the ancient classics. Cities and the ruins of cities such as Antioch, Ephesus and Troy, which William Lithgow had visited, were on his route. Eventually he would have sailed through the Sea of Marmara and arrived in the Golden Horn, alighting in the great capital of the Ottoman Empire, Constantinople.

CHAPTER FIVE

Aleppo

Achille de Harlay, Baron de Sancy

Constantinople was an obvious port of call for George Strachan. His intention in going there may have been to obtain a licence to travel throughout the Ottoman Empire, just as Pietro Della Valle was to do several months later. But he may have chosen the capital city for another reason. He needed access to a cosmopolitan society and learned institutions to improve his language skills. Just as important for the impecunious Scotsman, he may have thought that he could obtain suitable employment while there. However, it offered another benefit. Resident in the city was someone he knew who would offer him hospitality. The French ambassador to the Sublime Porte, Achille de Harlay, Baron de Sancy, had met Strachan when they both attended the court of Henri IV in Paris. This is known from the account of Pietro Della Valle who was aware of Strachan's stay with the ambassador. The Scotsman had left shortly before the Roman arrived to stay with de Sancy. In his journal, he described the French ambassador as the Scotsman's friend (Della Valle 1664: vol. 2, 437).

Strachan and de Sancy had a number of things in common. As well as their experience at the French court, they were polyglot scholars with an interest in oriental languages. Previously, both had strong connections to the Catholic Church, especially the Society of Jesus. Harlay had been bishop-elect of Lavaur before giving up the ecclesiastical life in 1601, when he inherited the family title on the death of his elder brother. His appointment as ambassador came in 1611 following the assassination of the king: the event that ended Strachan's chances of preferment in Paris and led him to abandon the royal court. The queen regent, Marie de Medici, gave Harlay, as ambassador, a specific remit to provide protection for Jesuits working in the Ottoman Empire who were under attack from Muslim fanatics (Goyeau 1910). There can be little doubt, as Della Valle wrote later, that Harlay was pleased to see his old court acquaintance (Della Valle 1664: vol. 2, 437). His position as ambassador was not an easy one and Strachan's convivial company would have been welcome.

It could not have taken Strachan long to realise that Constantinople afforded him few opportunities of employment or restoring his finances. Just as Ottoman subjects were required to stay in a reserved district while residing in Venice, foreigners in Constantinople were restricted in where they could live. The district of Pera lay outside the walls on the other side of the Golden Horn separate from the main part of the city. Under the Byzantine emperors, it had been the fortified trading station of the Genoese, but the Ottomans had removed its defensive walls and directed all visitors from the West to live there. This proximity did not lead to an atmosphere of camaraderie among Westerners. Their attitude to each other was a replication of the political rivalries of their homelands. The Venetians formed the largest community and saw other Europeans mainly as competitors in trade rather than allies (Ari 2004: 39–40). Under these circumstances, while staying with the ambassador Strachan would have had limited opportunities to develop the acquaintance of any visitors of nationalities other than the French, despite probably having friends in common.

Access to the Ottoman court would have been even more out of his reach. Earlier Strachan had tried with no success to obtain the position of diplomatic representative of the papacy to the British Crown (Chapter 3) and he may have thought of trying to obtain an equivalent position with the Ottoman court for one of the European powers, but it would have been impossible. Protocol dictated that the ambassadors could only approach the court through the services of a dragoman. To the sultan, the ambassadors were infidels whom the court considered to be little more than the representatives of merchants (Ari 2004: 47). The Ottoman opinion was not entirely unreasonable, given that the English ambassador was paid by the directors of the Levant Company who chose the incumbent. Only after 1620 did King James assume the responsibility of appointing the ambassador (*Calendar of State Papers* 1857: vol. XII, 248). Dragomans were the officially recognised state and diplomatic interpreters in the Ottoman Empire. They were issued with a formal certificate (*berat*) by the sultan, without which they could not operate (Ari 2004: 43). There were two levels of *berat*: one given to dragomans of the imperial divan and one for those employed by European powers. Some dragomans held both levels, but usually European ambassadors had to employ the services of a dragoman, who dealt with imperial dragomans, who in turn dealt with court officials, who were the only ones who could deal directly with the sultan. The profession of dragoman involved more than mere language interpretation. They fulfilled a range of diplomatic, consular and commercial duties. Initially drawn from all language and faith groups within the empire, over time they formed dynasties and grew to have considerable power and influence which they guarded jealously (Gürkan 2015: 110–11). Despite justifiable misgivings regarding their loyalties, the Western powers had to employ at least one of these dragoman families as their representative at court, and in their other dealings with Turks (Rothman 2009: 771–800). In 1585 the Venetian *bailo*, Lorenzo Bernardo, complained to the *signoria* that while negotiating in Constantinople he had to 'speak with others' tongues, hear with others' ears, negotiate with others' brains' (Rothman 2009: 115). However, Venetian complaints on this matter were somewhat hypocritical since they operated an even more restrictive system for Muslim traders in

Venice. In 1587 a number of Turkish merchants in the city complained to the *signoria* that the sole dragoman of the Venetian Chancellery, Michel Membré, was overcharging on their fees, and that more dragomans were needed to deal with the volume of business that they conducted. The dragoman's influence with officials and the profits he made were too great for the *signoria* to respond immediately, and it was only on Membré's death that two dragomans were appointed to replace him (Dursteler 2002: 113).

In Constantinople de Sancy employed as secretary a dragoman named Deays who was secretly prejudiced against him. Deays did little to improve his employer's standing with the Ottoman authorities who resented the ambassador's interference in their actions against the Jesuits. Three years after Strachan's visit, officials of the Sublime Porte arrested de Sancy along with some of his staff and tortured them in the belief that the French had been complicit in organising the escape of a European who was a state prisoner. Catherine de Medici was forced to recall her ambassador for his own safety (Goyeau 1910).

In light of the difficult atmosphere in which he was required to work, it is understandable that de Sancy was reluctant to see his friend leave, and continued to entertain him for several months. During this time, Strachan would have been able to improve his knowledge of Turkish, Arabic and possibly Farsi. It is not known whether he had any knowledge of these languages before he set out on his eastern pilgrimage. In Rome there were priests of the Churches of the Eastern Rites whose native language was Arabic and Strachan may have met them (Dellavida 1956: 45). Also, while at the University of Paris, he may have met Etienne Hubert, a physician who had lived in North Africa for several years and taught classes in Arabic for a short period (Lefranc 1893: 383). However, it is most likely that his first exposure to the Turkish and Arabic languages was with his fellow passengers during the sea voyage from Venice. In each case he would have had little opportunity to engage with the extensive literature of these languages.

Strachan had his first known opportunity to study Arabic and Turkish texts on gaining access to the ambassador's personal library. It would have been a point of intellectual pride for de Sancy to show the Scotsman his collection of books. De Sancy was a noted collector of rare oriental manuscripts which he took with him on his return to France. Many are now in the Bibliothèque Nationale in Paris (Dellavida 1956: 42). Access to these books would have allowed Strachan to become familiar with Arabic script which, with some variation, was used also in Turkish and Persian writings. Learning the mechanics of reading oriental languages was only a start. The Turkish spoken by the ordinary people on the streets of Constantinople was very different from that used at court and in Ottoman literature. Any knowledge of demotic Turkish that Strachan had gained by that stage would have helped little in his exploration of de Sancy's library. There, he would have found that Ottoman court Turkish was largely composed of Arabic and Persian words with fewer than a fifth being Turkish. This had come about due to the high regard among aristocratic Turks for the literature of the two older cultures. It may have been at this point in his travels that Strachan came to realise that his efforts should be directed towards gaining a thorough understanding of Arabic and Farsi, in order to fully appreciate the literatures of the Middle East. Later, when he was able to afford to do so, he began collecting manuscripts in both languages as well as a small number of texts in Turkish. Unlike de Sancy, he did not spend his money on acquiring rare ancient books, but usually paid for copies to be made. By this means, he acquired a much larger library with a wider range of subjects and authors than the French ambassador was able to accumulate.

Jabal Lubnān

According to Della Valle, Strachan left Constantinople shortly before the Roman arrived in August 1614 (Della Valle 1664: vol. 2, 437). His next major destination was the Arab city of Aleppo in Syria, a trading centre on the Silk and Spice Roads. It contained a large cosmopolitan population with many Europeans, and Strachan probably hoped that it would offer him opportunities that were denied him in the heavily regulated Ottoman capital. As he almost certainly had done on his journey from Jerusalem to Constantinople, he stopped at a number of points on the way in his quest to explore different parts of the Middle East. During his stay with the French ambassador, the two friends would have discussed the political situation in the Ottoman lands and what Strachan learned may have influenced his decision to visit Mount Lebanon after leaving Constantinople.

The Ottoman sultans, who had succeeded Suleiman the Magnificent, were not as capable or as fortunate as he had been. He had expanded his empire and controlled it successfully through governors who were responsible for their provinces, keeping order and collecting taxes on his behalf. The next three sultans gradually lost this degree of power and control. In the East the Persians regained territory they had been forced to cede to Suleiman. Also, they had expanded their empire further at Ottoman expense by capturing some of the Christian kingdoms of the Caucasus that had fallen under Turkish control. Later, in 1623, the Persian shah added to the humiliation he was heaping on the Ottomans by capturing Baghdad. Within the weakened Ottoman Empire, provincial governors began acting independently from the central government in Constantinople. This was particularly true of the Druze governor of Eyalet Jabal Lubnān (Province of Mount Lebanon). Its hereditary emirs belonged to the Ma'anid family and had forced the sultans to award them *iltizam* (the concession of retaining all taxes raised in return for providing military protection for the province). With their increased wealth, the Ma'anids began to act autonomously, and extended their territory to include the coastal plain of the Levant from Sidon in the north to Safad in the south and eastwards to Palmyra in the western desert of Syria, an area significantly greater than modern Lebanon (Nisan 2002: 96).

In 1608 the ruling emir, Fakhr al-Din II, signed a commercial treaty with Duke Ferdinand of Tuscany and welcomed Europeans for the purpose of trade. The Florentines set up a consulate and the French established a trading depot in Sidon. The northern part of the mountains of Lebanon was inhabited almost exclusively by Maronite Christians, and Fakhr al-Din not only opened up the whole of the province to them but welcomed Latin Christian missionaries (Gordon 2013). In 1619 he further showed his tolerant attitude to European Christians by giving the Franciscan Order custody of some of the holy sites in Galilee: Mount Tabor, the believed site of Christ's Transfiguration, and the grotto of the Annunciation in Nazareth (Heberman 1905–14: vol. 15).

When Strachan visited Mount Lebanon in 1614 he may have believed that it would provide a better environment in which to interact with native speakers of Arabic, and possibly allow him to earn enough to support himself while he studied their language. However, he did not stay long. It was his misfortune that his visit coincided with a period of instability in the region. Sultan Ahmed I (the Fortunate) had brought forces to bear on his provincial governor to bring him to heel. The emir had gone into exile in Florence in 1613 for his own safety. At the time of Strachan's visit he had just returned to confront again the imperial forces. His attempt to regain control was unsuccessful, and he was forced into exile in 1615, this time in Naples. Although the emir returned the following year and succeeded in regaining his fiefdom, the situation in Lebanon in 1614 was too unstable for Strachan to stay and he continued his journey to Aleppo.

Third City of the Empire

Aleppo was the third-largest city in the Ottoman Empire after Constantinople and Cairo. The Ottomans ensured that this rich city remained securely under their control with its *aga* (governor) being appointed directly by the sultan. It was of very ancient foundation, and owed its prosperity to its location at the crossroads of major trade routes. Raw silk and finished silk fabrics produced in China, India and Iran together with fine cotton fabrics and spices from India and further east came by camel caravans across the desert or by sea to Basra. Goods were then transported by way of Baghdad and Damascus onwards to Aleppo where they were sold to traders, some of whom took their purchases north to Anatolia while others travelled west to the Mediterranean coast, where the merchandise was transshipped, mainly on Venetian vessels, from Tripoli and other Levantine ports. Aleppo was central to this network, being located halfway between the river Euphrates and the Mediterranean. It lay at an ideal point for merchants from the East and West to trade.

The city catered for them by providing khans (fortified inns) such as Khan al-Shouneh where the merchants could lodge and store their goods safely. (This Khan survived intact until the Syrian civil war in the twenty-first century when it was largely destroyed along with many other buildings of the old city.) The twenty English merchants, who were there when Strachan arrived, had set up their base in Khan al-Burghul. By 1680 their numbers had more than doubled, and they moved into the larger and more imposing Khan al-Gumru (Mills 2017: 274). To facilitate trading, the city provided souks (large covered marketplaces) in which the merchants conducted their business. Souk al-Madina was the largest and was arranged into areas specific to individual merchandise, ranging from expensive items such as silks and spices to the mundane soaps and copper pots. Another prominent marketplace was Souk al-Zirb which specialised in supplying the needs of the Bedouin. As nomads the Bedouin tribes were very self-reliant, weaving the black wool cloth for their tents using the fleeces from their herds of sheep and goats and making their own horse and camel harnesses. However, they relied on the souks to provide the necessary worked metal pieces. Other needs included items of convenience and luxury, such as coffee pots and cups, but unlike city dwellers all such items had to be easily portable as well as durable. The souk in Aleppo was well placed to provide such items.

European governments had set up consulates to support their many nationals who were trading in the city. Venice was the first to do so shortly after the Ottoman conquest of the city from the Mamluks in 1516. At that time the city had a population of 50,000, but it soon grew to 90,000, eclipsing Damascus as the dominant trading centre on the Silk Road to the Mediterranean ports. The Venetian consulate was followed by those of France and England. The individual English merchants who traded in Aleppo had been present for a number of decades prior to the establishment of their consulate in 1583, and some of them remained independent and did not trade under the auspices of the Levant Company of London until 1620. In 1613, two years before Strachan arrived in the city, the United Provinces of the Netherlands set up their consulate to support their merchants who were attempting to expand on the trade with the East that the Dutch East India Company (VOC, Vereenigde Oost-Indische Compagnie) was successfully establishing by sea around the Cape of Good Hope.

The concentration of European traders made the extension of the Venetian postal system to Aleppo worthwhile. The Venetians allowed others to benefit from the service since it not only helped defray its running costs but gave them the advantage of being able to control and, in some cases, spy on others' communications (Dursteler 2009: 601–23). George Strachan used this service while resident in Aleppo in 1615. He wrote a letter to Christophe Dupuy proposing that he recommend that the king of France, Louis XIII, should collect Arabic manuscripts in the same manner as his predecessors had done with ancient Latin and Greek classical literature. The reason he gave was that such a collection would help in '*la propagation de la foy*', 'spreading the faith'. In it he also wrote that he intended to remain in the East for a further two years before returning to Europe (BN

(Dupuy) Ms. 712, f.108). Dupuy was a Carthusian monk who, at the time, was a member of Cardinal Du Perron's household and had befriended Strachan while they attended Henri IV's court. The survival of the letter suggests that Strachan used the postal system to write to other friends in Europe, such as Thomas Dempster. Its contents also imply that Strachan thought of his journey to the East as partly missionary in nature, and that his study of Eastern languages and literature would be of help to the Church.

In the early seventeenth century Aleppo was a truly international city. Merchants came not only from Europe but from India, Iran, the countries of the Caucasus and different parts of the Ottoman Empire to trade. With this plethora of peoples came a mixture of languages and religions. Conducting business in each merchant's own language was impossible, and the city used the lingua franca which was in common use around the Eastern Mediterranean lands. The Italian city states, especially Venice, had dominated trade with the East for centuries and a simplified form of their language had developed for business use. As well as being spoken in negotiations, it was used in written contracts among the diverse national groups trading in the city (Dursteler 2012: 47–77). Diversity of language was accompanied by diversity of religion. There were Muslims of all sects, Christians – Orthodox and Latin, Catholic and Protestant – as well as members of the religions of India. These were relative newcomers when compared with the established communities of Syriac Christians and Jews who had been resident in the city for centuries before the birth of Mahomet. Aleppo was one of the most cosmopolitan and tolerant cities in the Ottoman Empire; however, it was not without underlying tensions. For the most part, the different communities lived apart and dealt with each other only as business required. The pragmatism of the people can be summed up by a proverb used by merchants in Aleppo – 'if you do business with a dog, kindly call him "Sir"' (Mansel 2016). Under these circumstances, on arrival Strachan would have had little choice but to seek out members of the European community.

As well as the resident merchants, Catholic religious orders were attracted to Aleppo. The longest established were the Franciscans in their house, Convento di Terra Sancta, which still survives as the monastery of Em Ram (Roncaglia 1956: 145–53). Later, in the 1620s, Carmelites, Capuchins and Jesuits arrived, but when Strachan came the Franciscan convent was the only representative of Catholicism in the city. The previous year, while staying with the Franciscans in Jerusalem, Strachan would have been informed of the locations of all the Franciscan houses in the Middle East and have known of their monastery in Aleppo. The Franciscan convent had been established in 1238 only eight years after that of the mother house in Jerusalem, which underlines the importance of the city, due not only to its trade but to its large Syriac Christian population (Golubovich 1906: 114). It would have been in keeping with his known behaviour for

Strachan to have sought out the hospitality of the Franciscans and have arrived bearing, if not a letter of introduction from Jerusalem, at least news of their Franciscan brothers in the mother convent. In turn his Franciscan hosts would have been able to inform him of the European Catholics who lived in Aleppo, and it may have been through them that Strachan learned of the presence of an acquaintance.

In his journal Della Valle writes that the Scotsman had an unnamed friend in Aleppo whom he describes as a Flemish doctor (Della Valle 1664: vol. 1, 579). It is possible that Strachan had befriended the doctor in Europe. Much of his time in Europe had been spent at numerous colleges which had important medical schools. The entries in his album amicorum include a number written by physicians who were either members of the college teaching staff or individual medical practitioners (Strachan's *Album Amicorum*: 23, 27 passim).

Even if the Fleming and the Scotsman had not previously met, they would have had mutual acquaintances, a circumstance that Strachan would have used as a means of introducing himself. The doctor was to prove important in helping him during his time in Aleppo. Strachan also would have taken the opportunity during his first few months there of making himself known to other members of the European community. Having travelled through much of Europe and being fluent in a number of languages, he would have found many willing to converse with him and hear his news especially when they realised that he had not come as a trading competitor. The connections he made at this time were to be important to him later, especially his contacts with the English merchants. Despite the fact that he had not come as a merchant and that his principal reason for travel was to conduct scholarly research, he was about to become involved in the Silk Road in a way that must have seemed remarkable to the merchants in Aleppo, and certainly was unique to Europeans of the time. The rest of Strachan's life was spent involved in various capacities in this ancient commerce, but at the time of his arrival in the first half of the seventeenth century it was undergoing major changes in the way it operated.

The Silk Road

The trade in silk was large and profitable but fiercely competitive. Merchants from three continents contested with each other for a share of the luxury markets in silk and spices.

From the earliest times, exports from China to the Middle East, North Africa and Europe had been taken overland through Central Asia. From there some merchants crossed the Himalayas to trade in India, Afghanistan and Persia while others continued through the Tatar regions and further west into Europe. In India merchants also traded in other luxuries, such as fine cotton fabrics and spices. Some spices were grown in southern India, but others such as nutmeg grew only in the Maluku Islands, part of modern

Indonesia. These spices were brought by sea to India and became part of the general trade.

Merchants from Egypt sailed across the Arabian Sea to ports along the west coast of India to bring back silks, cottons and spices. When the Portuguese arrived in these waters by sailing around the Cape of Good Hope at the end of the fifteenth century, they created another important route to Europe for these commodities. They were followed by the Dutch, French and English who became fierce trading rivals. From this it can be seen that the Silk Road cannot be described as a single linear entity, but rather as a network of routes along which the merchandise was moved. If, for reasons of war or politics, parts of the network were rendered impassable, goods would be brought westwards along alternative routes. During the period when the Egyptian Mamluks and Ottoman Turks contested for supremacy of the Arab world, each of these powers fostered a major trade route through the territory they controlled. When, in 1517, the Ottomans prevailed over the Mamluks, they gave preference to their trade route through Syria to the Levant at the expense of the Egyptian route using the Red Sea. At about the same time, the sea route from India to Basra in Ottoman-held Syria fell under the control of the Portuguese. They built forts at strategic points around the Arabian Peninsula. Those in the Straits of Hormuz were of particular importance since they controlled access to the Persian Gulf. As a consequence, the overland route through Persia to Baghdad and onto Aleppo became more important. This route's dominance increased from the beginning of the seventeenth century when Shah Abbas I of Persia transferred silk manufacturing from the Caucasus to build a major production centre of high-quality silk fabrics in his new capital of Isfahan (Chapter 9). The shah took great pains to ensure that he controlled as much of this commerce in silk as possible, and merchants found that the safest and least taxing way to deal was to take the route westward from Persia through the Syrian Desert to Aleppo. Strachan's journey to Aleppo had taken him to what had become the principal point of convergence of trade routes in the Ottoman Empire.

Emir Feyyād Abū Rīsha

The trade involved caravans from Persia crossing the Great Syrian Desert in order to reach Aleppo. The desert covers more than 200,000 square miles and straddles what at the time of Strachan's visit were the territories of the Safavids and Ottomans, but today includes modern Syria and Iraq as well as western Iran. The river Tigris formed the boundary between the empires. West of the river in the Ottoman lands, the desert was ruled by a Bedouin emir of the Anazzah tribe. In the days of Suleiman the Magnificent, as was true of all provincial governors, the emir was subject to the sultan but, as with the Druze emir of Mount Lebanon, the Bedouin emirs of the Syrian Desert had gained independence from the Sublime

Porte and been awarded *iltizam*. This concession led to the emir, Feyyād Abū Rīsha, becoming extremely wealthy. His reputation was such that the European merchants based in Aleppo referred to him as the 'King of the Arabs' (Oppenheim 1939: 305–12). Emir Feyyād had managed to achieve this dominance while retaining the goodwill of the sultan in Constantinople. Although the western part of the Syrian Desert was nominally part of the Ottoman Empire, the sultan had to treat the emir as an ally rather than a subject. The Bedouins were extremely useful to him due to their ability to control the land along the border of the Ottoman's enemy – Safavid Iran. Also, Emir Feyyād had come to the sultan's aid in suppressing the Druze emir, Fakhr al-Din II, during his rebellion in 1613. Feyyād had good reasons to side with the sultan in this conflict. Fakhr al-Din had built a fortress at the ruined city of Palmyra which encroached on Feyyād's territory, and in support of his rebellion he also had enlisted the help of Feyyād's brother and cousin who, at the time, were attempting to supplant their emir as leader of the tribe. Later, in 1621, the cousin, Mudlij, succeeded in overthrowing Feyyād and becoming emir (Oppenheim 1939: 312). By showing effective support for the sultan in 1613, Feyyād had gained a unique position within the Ottoman Empire. He remained on good terms with the sultan despite being completely independent of his authority.

Early in 1615, this powerful Bedouin emir set up his encampment outside Aleppo and let it be known in the city that he wished to employ a personal physician. George Strachan presented himself for the position and was accepted (Della Valle 1664: vol. 2, 718). This outcome appears extraordinary on a number of grounds. Strachan was an infidel (*kafir*) in the eyes of the Sunni emir: in addition, he was not a doctor and Aleppo had many qualified Muslim doctors including Arabs whom the emir would have been expected to prefer. The surprising development may be explained by the particulars of the offer of employment being made. As personal physician, it was a requirement that he attend Feyyād wherever he went. The capital of his province was the town of Āna which was located on an easily defended position on the Euphrates controlling the central trade route through the desert. However, as a man of the desert (the meaning of the term Bedouin, in Arabic '*badawī*') the emir rarely stayed there. His nomadic lifestyle required him to spend his time travelling around his 'kingdom' leading the tribe's herds to new grazing and collecting taxes from the travellers and the many small settlements in his region. The emir and his household, which consisted of many hundreds of followers, lived in tented encampments as a rule. Such a life would not have been to the taste of the city-dwelling doctors of Aleppo. Added to this was the opinion held by non-Bedouin Arabs (*ḥāḍir*) that the men of the desert were by nature savages, a view they had formed through many years of suffering at their hands in raids on settlements. Even large cities were not immune from their attacks. A little over a century earlier, in 1480, a Bedouin army had succeeded in sacking

Jerusalem. Its governor only narrowly escaped with his life while many of the city's population were killed (Montefiore 2011: 288). The Bedouin continued to deserve their fearsome reputation. In the last quarter of the eighteenth century when Aleppo was much reduced in the commercial importance it had held a century earlier, a Bedouin army raided the city, looting and killing (Faroqhi et al. 1997: 788). Muslim doctors in Aleppo would have had grave misgivings about accepting a position in such a community where life would not be easy and the consequences of the emir's displeasure could be fatal.

There would have been few other candidates available to the emir, since it is almost certain that the only European doctor in the city was Strachan's Flemish friend who already earned his living by tending to the merchants from the West. The Franciscan friars would have been able to provide limited medical treatment in their monastery, but it would have been impossible for any of them to take up the position of physician to the emir. Maintaining their health while stationed in the East was a problem for European merchants. Those from Mediterranean Europe may have had some immunity to diseases prevalent in warm climates but those from the north had little natural defence and were susceptible to a range of exotic illnesses. Della Valle wrote in his journal of numerous occasions when he and members of his party suffered from outbreaks of disease. His wife, of Arab and Armenian descent, died in Persia from the plague and he wrote that later in his travels his friend, George Strachan himself, succumbed to malaria (Della Valle 1664: vol. 2, 158). It is known from its records that the English East India Company had difficulty in finding suitably qualified doctors willing and capable of service in the East (Elgood 2010: 393). Despite misgivings by some of the company officials, they were later to use the unqualified Strachan in that capacity to look after their employees in Isfahan in Persia. It is likely, therefore, that there would have been few contenders for the position of personal physician to the emir.

There is evidence that Strachan's medical knowledge was greater than might have been expected from a classical poet and scholar. As has already been noted, he had gained much of his education at colleges that housed medical schools; the universities of Lescar and Montpellier were particularly renowned for the study of medicine. His album amicorum shows that Strachan counted a number of their students and professors among his friends. By association with them, it is likely that he gained a theoretical knowledge of medical matters. In addition, he may have attended some of their lectures and demonstrations and gained a more practical knowledge, while never following a formal course in a medical faculty. In the few months he spent in Aleppo in 1614–15, Strachan also may have further improved on any medical knowledge he had by acting as assistant to his friend. The Flemish doctor would have had more patients than he could have dealt with, and at times may have welcomed his help. In return Strachan would have benefited from some much-needed income. Given this background it can be seen that Strachan's application for the position as physician to the emir was not as foolish as it might at first seem.

From Della Valle's journal we learn that when Strachan first visited Feyyād, the emir was suffering from some minor ailments (Della Valle 1664: vol. 2, 718). Prior to his first consultation with Feyyād, Strachan's Flemish friend had given him a small stock of drugs and herbal medicines and had instructed him in their use. With this limited medical knowledge, Strachan was fortunate in being able to relieve the emir of his discomfort. His initial success helped create a good impression, but in itself would not have been sufficient to convince the emir that he should be given the important position of his personal physician. Strachan's strength of personality must have played an important part in the decision. Throughout his career he had been able to make friends easily and had spent much of his adult life at a number of important courts in Europe – the Papal and French as well as the time he spent as personal mathematician in the household of the Duke of Guise. He was familiar with the etiquette required in such company. In addition he was always conscious of his own noble birth which made it natural for him to relate on equal terms to members of the nobility. Emir Feyyād appears to have been much impressed by the competent and confident Strachan. Della Valle wrote in his journal that in his discussions with his patient the Scotsman was sufficiently bold to advise the Bedouin ruler that his health would be improved greatly by restricting his sexual encounters to one wife (Della Valle 1664: vol. 2, 718). It is difficult to imagine a less confident individual daring to venture onto such a subject. The emir must have been taken aback. By his action, Strachan gained an important ally in the person of the chief wife of the emir's harem, who was delighted to hear that her husband's access to her rivals should be restricted in this way. In offering such advice, Strachan was demonstrating knowledge of a medical theory current in Europe.

Doctors believed that illness was influenced by the state of the patient's blood. Heated blood, evident when a patient was running a temperature, exacerbated the illness while cool blood helped to ameliorate it. Sexual activity was believed to overheat the blood, and therefore should be avoided or, since this was impractical in normal situations, restricted as much as possible. These ideas were developments on the work of Marsilio Ficino, the fifteenth-century Italian humanist philosopher. In his *De Vita* published in 1489, he argued that sexual intercourse involved the loss of male fluids that were equivalent to blood, and, as blood loss was injurious to health, likewise intercourse was injurious to men's health (Matteoni 2009: 109). Della Valle does not record whether the emir followed Strachan's advice on this matter, but the Scotsman retained the good favour of the emir's wife for the duration of his time with the Bedouins.

It is possible that George Strachan may have had another reason for offering the emir advice on sexual continence. He may have had specific concern regarding the emir's exposure to sexually transmitted diseases.

This would have been the case if one of the emir's 'minor ailments' was syphilis, a disease which appears to have arrived in Europe in the late fifteenth century. It is believed to have been introduced from the New World by early Portuguese and Spanish explorers. By the end of the following century, it had reached epidemic proportions, spreading through Europe, North Africa and Asia. To the Arabs it was known as the Franks' disease. In Europe physicians had given it the same name in Latin – *morbus gallicus*. The disease and its symptoms would have been known to Strachan as a matter of course, but he must also have had a theoretical knowledge through the work of Girolamo Fracastoro, a Veronese polymath, physician, poet and classical scholar. In 1530 Fracastoro wrote a long epic poem, '*Syphilis sive morbus gallicus*', which received international acclaim, and was known to scholars in Strachan's time at university. It told the Greek myth of the shepherd boy, Syphilis, who insulted the god Apollo and was punished with a horrible disease. In his poem Fracastoro described the symptoms of the new disease and suggested the use of mercury (an extremely dangerous procedure) and guaiacum in its treatment. Guaiacum was one of a number of medicinal plants introduced to Europe from the New World. These new plants were used to treat different conditions. Another introduction from the Americas was cinchona, also known as Jesuit's bark. The Jesuits learned of its medicinal properties from Native Americans and introduced it to Europe as an effective treatment for the symptoms of malaria. Guaiacum was not as effective against syphilis, however, having only ameliorative properties. The great interest which Fracastoro's work generated was due to the general concern regarding the rapid spread of syphilis. As a result and in order to satisfy the demands of his fellow physicians, the poet was required to produce a full medical treatise on the same subject in 1546 – *De Contagione et Contagiosis Morbis* (Beretta 2009: 129–54). Strachan may not have read the treatise but the poem certainly would have been known to him. Also, it is likely that he had a supply of cinchona and guaiacum in the medical chest he received from his Flemish doctor friend. If the emir was suffering from the early stages of the disease, it would provide another reason for selecting Strachan as his personal physician. A Frankish doctor, as the emir would have considered Strachan, to treat the Frankish disease would have been a logical choice.

On his appointment, Strachan received a payment from the emir which Della Valle described as generous (Della Valle 1664: vol. 2, 718). With it Strachan returned to Aleppo and bought a number of items which he felt would be of help to him in his new position. As well as clothes suitable for travelling in the desert he provided himself with medical books, no doubt purchased from his Flemish friend. With these few possessions and his chest of medicines, George Strachan set out on his great adventure into the Syrian Desert in the company of the Bedouin emir's tribe and dressed as one of them.

CHAPTER SIX

Mohammed Çelebi

Desert Life

The first account we have of Strachan in his new role was written a year after he entered the emir's service and was given by Pietro Della Valle. The Roman nobleman and his party had joined a caravan of merchants in Aleppo to cross the desert to Baghdad. The caravan had stopped at Āna, the chief city of the Anazzah Bedouins, where Strachan had the opportunity to learn about Bedouin life. In a letter home written a month later, he explained that he had been in the realm of Emir Feyyād and went on to write:

> [L]iving today with the emir Feiad is one of our Christians, a gentleman of the Scottish nation, called George Strachan, a Catholic, and an educated and much respected man . . . he won such a reputation with him [the emir], and such good favour, that he is now Master of the Rod, and the most favoured at court, as well as having acquired the money and many conveniences he needed . . . So he is very well thought of by everyone; and when one says the name Strachan in the desert, one need say nothing more. (Della Valle 1664: vol. 1, 579: Bull 1989: 98)

Della Valle and Strachan were to become good friends, and his later comments may be coloured by this relationship but, when he wrote his letter in December 1616, he had never met the Scot. It is clear that after being in the desert for a year, Strachan's material circumstances had improved greatly. The post of 'Master of the Rod' can best be explained as an honorary role in the emir's entourage, and underlines Della Valle's comments on the respect in which he was held among the Arabs. The emir valued his services as a doctor but this cannot explain the high regard that Della Valle describes. Feyyād would also have been impressed by the ease with which the Scot conducted himself in his presence. Strachan was born of a noble family and, as well as his childhood instruction in the manners of a gentleman, he had learned how to behave in the presence of royalty and nobility during his life at European courts. The emir enjoyed his company, which must have contrasted markedly with the way that he dealt with male members of his own family. Two years earlier his brother and cousin had risen in revolt against him, and the loyalty of other members of the family may have raised questions in his mind. At the very least the atmosphere between the emir and some of his relations would have been guarded. Feyyād would have been able to relax in Strachan's company, knowing that the Scot would not overstep the bounds of protocol. He was intelligent, well-educated and good company but, most importantly for Feyyād, he had no prospect of advancement other than through the emir's favour. It would not have been in Strachan's interest to side with any faction within the tribe.

Bedouin life was comfortable despite being nomadic and for Strachan it had many attractions. The Anazzah tribe was wealthy due to its control of the desert caravan routes, but their style of life revolved around their herds of animals. The traditional Bedouin measure of wealth lay in the size and quality of their flocks and herds. Dowries were usually calculated in numbers of animals rather than money. Sheep, goats, camels, donkeys and especially horses were highly valued and their care was a priority for their owners. As desert dwellers, this meant a constant search for new pastures. Their land extended to more than 100,000 square miles of the Syrian Desert but the town of Āna on the river Euphrates was the only permanent settlement the Bedouins occupied. For the most part, they lived in their black tented encampments and rarely stayed more than two weeks in any place. The tribe was a community of families that by necessity spread over wide areas in order to avoid exhausting pasture too quickly, and when they moved on it was at the pace of the slowest animals with their flocks grazing as they went. Their peregrinations were not random, but took pre-determined routes with progress dictated by the seasons. Strachan described his time with the Bedouin to Della Valle when they formed their friendship in Persia. In his letter of 20 November 1622 the Roman wrote that his friend found:

> the utmost relish [in his life in the desert] . . . due not merely to the pleasure of constant wandering (at a gentle pace, indeed, which caused no fatigue) but also to the noble pastime afforded by sport in various forms to which the chiefs are given; and to that generous manner of living in absolute freedom to which those people are habituated, neither hemmed in within town walls, nor subject to the rules of anyone except of the prince when he is present.

The Bedouins did not compromise their love of freedom even when they lived in Āna. The town sprawled for five miles along either side of the river Euphrates. Its houses were widely spaced and set in large orchards. The town was unconfined by any walls and relied on the surrounding desert for protection. The sports that Della Valle mentioned as being available to the elite of the tribe centred on hunting, particularly falconry, but horse and camel racing were popular and any male member of the tribe could participate. As a European of gentle birth, Strachan would have been familiar with all of these pastimes (with the exception of camel racing) but most of his life had been spent as an impecunious scholar and he would have had little opportunity to indulge himself in such luxuries. It is not difficult to understand why his life in the desert appealed to him. Strachan seems to have enjoyed the 'absolute freedom' of movement of life as a Bedouin. Also, by being a privileged member of the emir's household, he had a degree of freedom which did not restrict him to the desert encampments. As well as days spent hunting in the desert or visiting Islamic scholars in their madrassas, he was able to take trips into Aleppo and Baghdad to restock his supply of medicines and purchase items for personal use. While there he searched for books to buy or have copied to augment his growing library. The emir was willing to permit Strachan to leave his territory on these visits but clearly wanted him to return to his service. Della Valle, in his letter of December 1616, tells of one such occasion:

> I myself can testify that a few months since, the emir, being in the desert not far from Aleppo, and Signor Strachan having come to the city on his own affairs, the former, who had intended to leave, stopped there waiting for him for more than a fortnight; and finally, when there was still a delay, he started indeed, but left behind one of his principal chiefs with more than a hundred horse to wait for Strachan, and to escort him safely through the Desert; or, it may be, to make sure that he did not slip away, should he be that way inclined! (Della Valle 1664: vol. 1, 581; Yule 1888: 315)

As Della Valle had written these words before he had met Strachan, it implies that general talk regarding the Scottish gentleman spoke not only of the high regard in which he was held by the emir but also of the ambiguity with which Feyyād viewed his physician. It is clear that even by this stage of his stay in the desert the emir was determined that Strachan should remain in his household.

Strachan had a number of reasons to extend his stay in Aleppo. His visits allowed him to renew his friendship with the Franciscans as well as with other members of the European community. The emir was beginning to put him under pressure to convert to Islam and while lodging at the Franciscan convent of Em Ram he was able to join in their daily religious services which would have strengthened his resolve to remain Christian. He appears also to have taken advantage of the Venetian postal service which operated from Aleppo, its eastern terminal, through Constantinople to Venice. This postal system was available to all Europeans. The Ottoman Empire had its own system for state communications, part of which involved using carrier pigeons for speed. The *aga* in Aleppo communicated in this way with the cities of the Mediterranean coast fifty miles away; however, this service was not available to Europeans (Moshenska 2016: 272). The information that Thomas Dempster included in his short biography must have been given in a letter from Strachan at this time and he was not the only friend in Europe to whom Strachan wrote using the Venetian service (BN (Dupuy) Ms. 712, f.108).

As well as associating with the European community, Strachan spent his time in Aleppo dealing with the Arab residents of the city, the *Ḥāḍir.* His status in the emir's household provided him with a prominence among them. In his letter of November 1622, Della Valle described their reaction to Strachan in this way:

> [W]hen he visits Aleppo, among the multitude of those who throng him and pay him court (to a greater degree even than is paid to doctors in Naples by their patients), the very Arabs do not distinguish him from a genuine Bedouin. (Della Valle 1664: vol. 1, 580; Yule 1888: 315)

Della Valle implies that the interest shown by the Arabs was caused by their wish to consult him as a doctor. This may have been true in part but the appearance of a wealthy Bedouin would have attracted attention for other reasons. The poor would have sought alms and the affluent would have been seeking favours at the emir's court. Also, Strachan had money to spend in a way that he rarely had before. He spent time researching books and, although only one survives that Strachan is known to have bought in Aleppo, there were others that have been lost (Appendix: uncatalogued Ms. No. 1 of 5). The purchase of books and the hiring of scribes involved visits to libraries and discussions with prominent citizens who possessed private book collections. Furthermore, his shopping expeditions through Souk al-Zirb to buy Bedouin finery would have generated an admiring audience as well as the attention of the merchants. It is clear that Strachan enjoyed his visits to Aleppo and Baghdad, but before too long he had to return to the desert and the emir's household.

Desert Caravans

Strachan's decision to offer himself as the personal physician to the emir had brought him significant benefits. In his letter of December 1616, Della Valle gives information on the lifestyle that Strachan shared with the Anazzah Bedouin. Much of their activity related to taxing the trading caravans which regularly crossed the great stretch of desert under Feyyād's control. This was the principal source of the emir's wealth.

The merchants who made up these caravans assembled at an agreed place and waited until their number had grown large enough to provide mutual protection crossing the desert. Thomas Coryate, the Jacobean court jester and traveller, self-described as a 'legstretcher', provided an account of the caravan he travelled with in 1614 from Ur in Mesopotamia to Isfahan. It was made up of 'two thousand camels, fifteen hundred horses, eighteen hundred mules and six thousand people' (Nicholl 1999: 3). Coryate's accounts of his journeys contain tall tales so it is to be expected that his description of the caravan is exaggerated but in essence it is not too different from that given by a more trustworthy source, that of Pietro Della Valle.

The caravan that the Roman nobleman joined in 1616 was typical of the many journeying between the Ottoman and Safavid Empires. It was composed of merchants and travellers, few of whom were European. Della Valle wrote that he and his party dressed *alla moresco* to avoid detection as Franks. Since he was fair-haired he had to shave his head as well as wear a turban. Other members of the caravan were aware of the Europeans' identity, but the disguise was needed to protect them from unwanted attention along the journey. They had joined the caravan where it was forming at a small town, Jibrin, six miles from Aleppo. All arrivals were checked by Ottoman officials for payment of levies and to ensure that no untaxed goods were being smuggled. After a week, by which time the caravan had grown to about 1,500 travellers with attendant camels, horses and donkeys, they set out, but travelled only seven miles before stopping at a village called Melluha. There, they had to wait for two days before the party of horsemen sent by Emir Feyyād arrived to inspect the goods and collect the duty to be paid. This location was stipulated by the Bedouin since they wanted to ensure that the caravan paid dues before crossing the desert, where it might have been attacked and robbed by bandits who were outwith the control of Emir Feyyād; although it appears that there was little likelihood of that happening in his domain. Della Valle speaks well of Feyyād, quoting merchants in the caravan who said that he was 'an upright man, and wins great obedience through his courtesy, and keeps his state completely clear of wicked people'. On other stages of the Roman's journey, there was real concern that the ruling authorities were unable to prevent rogue elements of the desert tribes from despoiling the caravans before or after the taxes had been paid. For that reason, caravans always travelled with an armed escort. Della Valle's caravan had 'janissaries and other formidable Turkish soldiers' including archers and swordsmen. The Roman was surprised to find, however, that among such a large group there were only eighty soldiers and merchants armed with harquebuses (Della Valle 1664: vol. 1, 560–5). In European armies, firearms – harquebuses and muskets – for the most part had replaced archers.

In his letter he gives no description of the taxes being collected at Melluha other than to say that the emir entrusted the task to 'one of his most intimate servants' and it took more than a day to complete. However, there are other accounts of the procedures used on the Iranian side of the desert (Chapter 9). The shah's instructions were, like Feyyād's, that the merchants stop at designated custom posts and display their wares, identifying those liable to taxation. The duty was then assessed by the officials and paid by the merchants. Afterwards, the Persian officials and the soldiers protecting them ransacked the merchants' belongings looking for any taxable goods that had been hidden. If they found any, they confiscated them and their value was shared evenly between the shah's treasury and the soldiers. Merchants would try to hide valuable goods to avoid payment and were ingenious at doing so. Equally the shah's officials became adept at discovering the hiding places and received compensation from the shah for their efforts. This ensured that they would be diligent in their work and that it was not in their interest to collude with the merchants to avoid the tax. Elsewhere, bribery was a normal occurrence in such transactions, as is shown by Della Valle's description of his tax assessment on arrival in Jerusalem which includes an account of his bribing the officials there (Chapter 4).

Emir Feyyād may have used similar means to reward his tax collectors to ensure their relative honesty, but tribal loyalty probably rendered such measures unnecessary. Della Valle's letter also shows that the tax wealth was spread throughout the tribe and not kept solely for the emir's use. When his caravan stopped at Āna his description of the population includes this passage:

> [They] are true Bedouins; but the most well-bred Bedouins in the world. Both in their dress and their appearance, they are not only dignified but also most fanciful, many of them wearing most curious silk dresses and bizarre abe or coats, long and stripped, and mostly black and white, or white and tawny, with a thousand curiosities in their tassels, knots, sashes, arms, headgear and other fripperies. (Della Valle 1664: vol. 1, 576: Bull 1989: 98)

It is clear that the Anazzah Bedouins were not only wealthy but displayed their wealth in silk clothes and personal ornamentation. During his stay with the emir Strachan shared in this wealth and in such personal displays of being 'a true Bedouin'.

Transformation to a Bedouin

In the two years that Strachan spent with the emir he transformed himself such that he looked, dressed and spoke like a Bedouin: to the *Ḥāḍir* he was indistinguishable from the other Bedouin princes. Later, when they met in Persia, Della Valle was greatly impressed by Strachan's knowledge of Arabic. In his letter dated 24 August 1619, he wrote: 'He is an excellent master of the Arabic tongue, and possesses, as well as has read, many and capital books.'

The Arab-speaking world in Strachan's time stretched from the Maghreb to Mesopotamia and in each region the dialect of the Arabic language used was distinct: all were mutually intelligible, although sometimes only with difficulty. In addition, all of these variations differed from the classical Arabic of the Qur'an which had been written in the seventh century of the Christian era. Muslim children were taught verses of the Qur'an by rote with much of the archaic language having to be explained by their teachers. Before he joined the Bedouin, Strachan had to deal with a number of different dialects of Arabic but then went on to master the form used by the Bedouin and through his teachers became proficient in the Arabic of the classical texts which he started to collect.

Seemingly idyllic though it was, Strachan's life in the desert was not without problems. A major issue regarding religion arose between him and his emir. When he finally met Pietro Della Valle in Persia, the two men spent many hours conversing about their background and experiences and it was during these discussions that Strachan explained the pressure that he was put under to convert to Islam. In his letter written in Combru (Gombroon, now Bandar Abbas, a port on the Persian Gulf) dated 20 November 1622, Della Valle wrote:

> They were also continually endeavouring to persuade him to become a Mahommedan, an endeavour which he rather fenced with and put off than met with a decided negative. And this, he says, he did, not so much to avoid offending the prince and his wife, as to show that his belief was not the result of haphazard; and that a change of faith should not be made for worldly ends but only if they should really convince him that their religion was better than his own. (Della Valle 1664: vol. 2, 719; Yule 1888: 317)

It would appear that Strachan attempted to maintain his favourable position with the emir through equivocation while adhering to his Christian faith. As a young man he had attended universities that were by confession Protestant while privately he held to Catholicism. Perhaps he thought of his sojourn with the Bedouin as simply another instance of minor deception for practical reasons. As events transpired, this was a dangerous assumption.

Raimondi's Arabic Gospels

A document exists which demonstrates the importance to him of his Christian faith and helps explain the efforts he made to retain it. The history of this book is crucial to understanding Strachan's later life since it appears he retained it until his death. Its survival is due to an officer in the Bengal Army of the British East India Company. Major William Yule served as assistant resident at the independent courts of Lucknow and Delhi under the command of General David Ochterlony. Both men lived in India for many years. The general was a striking individual who, as resident in Delhi, fully embraced the manners of the Mughal court, adopting Indian dress and taking thirteen wives whom he regularly paraded around the walls of the Red Fort, each on the back of her own personal elephant (Coleman 2004). Under the influence of this remarkable man, Major Yule developed a respect and admiration for the civilisation of his host country. In particular he took a scholarly interest in the Persian and Arabic languages and culture of the Mughal court. In the process he accrued a sizable library of books and manuscripts. On his death in 1839, his sons presented his complete collection to the British Library (until 1973 part of the British Museum) where it remains today. Included among the Arabic volumes is a translation of the New Testament which has fine wood-cut illustrations of biblical scenes by the Italian artist Antonio Tempesta. The book is folio-sized (12 by 19 inches) and was printed in 1591 by John Baptist Raimondi at his Oriental Press in Rome. On its end fly-leaf (which, since Arabic script runs right to left, is the start of the book) is handwritten:

> *Legit Georgius Strachanus Merniensis Scotus; Diebus viginti, horis succisivis; In desertis Chaldeae ad occidentem Babilionis (M?) et apud Faiathum in Regem Arabum anno Chri[sti] 1616 finivit die 19 Januarij. Summa Laus Trinitati individuae*

The text has been damaged and obscured in part; nevertheless, its meaning is clear:

> George Strachan of the Mearns, a Scotsman, read [this book] in twenty days during his leisure hours, in the deserts of Chaldaea, west of Babylon [Baghdad] in the realm of Feyyād, king of the Arabs, in the year of Christ 1616. He finished [it] on 19 January. High praise [be] to the Indivisible Trinity. (Dellavida 1956: 70)

Strachan's possession of this book gives an insight into his thinking when he joined the service of the emir. He must have obtained it in 1615 having purchased it or possibly received it as a gift from the Franciscans with whom he had been staying. Its printer, Raimondi, had set up his Oriental Press in the 1590s to produce books to be used on the Catholic Church's missions. The translation that Strachan owned is known variously as the Arabic, Alexandrian or Egyptian Vulgate, and was the first ever printing of the Gospels in Arabic. It has recently been argued that it was derived from a mid-fourteenth century translation by Coptic monks in an early Christian monastery (ancient Scetis) at Wādī al-Naṭrūn in the Libyan Desert, west of the Nile delta (Halft 2017: 301). Few of these books were sold in Europe: the majority went to missionary stations in the East. The missionaries in Aleppo would have received copies, and Strachan's Franciscan hosts, on learning of his plan to live in the desert with the Bedouins, are likely to have discussed with him the opportunity it presented for spreading the Christian gospel.

Given his avowed Catholicism and his long association with members of the Church's hierarchy, Strachan saw his purpose in journeying to the East as partly missionary in nature. The value of his converting a Bedouin emir to Christianity would have been enormous. Conversion of the ruling elite was the typical strategy employed by the Society of Jesus in its worldwide missionary activities. Having received the greater part of his early education from Jesuits, Strachan would have been familiar with their ways and would have found an edition of the gospels in Arabic extremely useful in any attempt at conversion. When he first joined the emir's household, he considered attempting this missionary task but he must have realised within a short time that it was impossible. Nevertheless, it appears he did try. Pietro Della Valle in his journal mentions that Strachan engaged in 'daily controversies . . . in which the part taken by him among those Mahommedans might be regarded as substantially preachings' (Della Valle 1664: vol. 2, 719; Yule 1888: 317).

For Strachan, however, the gift, if that is what it was, of the four Gospels would have served another purpose. From the start he would have known that, even while he was attempting to convert the Muslim, it was likely that pressure would be put on him to convert to Islam. Possibly his missionary friends expressed worries regarding the danger he was facing in simply being exposed to the 'false religion' of Islam. Western Christianity's confessional antipathy to Islam was not restricted to practising Muslims. The Catholic and Protestant Churches also viewed Christians living under Islamic rule with suspicion. The Christians of the Churches of the Eastern Rites were required to compromise their behaviour in order to comply with their Muslim overlords' governance and, in the eyes of the Western Churches, this meant that they were 'tainted with error'. The practical manual for witch-hunters, *Malleus Maleficarum*, '*The Hammer of Evil-doers*', a book much loved by King James VI/I, propounded this belief. Published in 1487, it remained the best-selling book after the Bible for 200 years (Guiley 2008: 223). Strachan would have been familiar with the text:

> There are two degrees of the Apostasy of perfidy. One consists in outward acts of infidelity, without the formation of any pact with the devil, as when one lives in the lands of the infidels and conforms his life to that of the Mohammedans. (Summers 1971: 76)

The text could hardly be more relevant to the life that Strachan was embarking upon in the Syrian Desert. He would have been fully aware that, in living with the Arabs, there was a danger that he could no longer adhere fully to the ordinances of the Catholic Church. The note he made in the flyleaf of his copy of the 'New Testament' shows that he read the book from cover to cover early in his stay with the Anazzah Bedouins. He achieved this in his spare time in less than three weeks. Clearly, Strachan had gained more than a practical understanding of Arabic, when he was first employed by Feyyād. However, he did not have complete mastery of the language, and working with a text with which he was already familiar in Latin would have helped. In addition it appears that he had a true affection for the text beyond its chrestomathic value. There is evidence (Chapter 12) that he retained the book for the rest of his life and treated it as a vade mecum. It is also likely that he used it as a primer for teaching Arabic to others, including at least one member of the English East India Company (Chapter 10) and several Carmelite friars in Isfahan (Chapter 11). Possession of this edition of the four gospels must have given him spiritual comfort and additional confidence to deal with the Islamic influences he would face in his role as personal physician to Emir Feyyād Abū Rīsha.

Conversion to Islam?

Notwithstanding Strachan's efforts to resist, there is no doubt that during his time in the desert the emir believed that the Scotsman had converted to Islam and treated him accordingly. Under these circumstances any attempt to return to Christianity would have been viewed by his host as apostasy and dealt with harshly. Della Valle makes clear that Feyyād viewed this arrangement as permanent and that it placed Strachan in great danger. In his letter dated 23 December 1616, Della Valle described Strachan's intentions in taking up the post as 'making a little purse and then retiring; for I can't think that kind of life for a continuance could be agreeable to one of us' (Della Valle 1664: vol. 1, 581). For most of his adult life, the Scotsman had sought the benefits that Emir Feyyād provided. He had looked for a wealthy sponsor as a source of sufficient income to allow him to pursue his academic studies, with no other arduous duties attached. The emir did this and more. The position that Strachan had originally obtained had been elevated to an extremely favourable one, but it tied him permanently to the service of Feyyād with the requirement that he convert to Islam. This had never been his intention.

The 'conversion' must have taken place in 1617, the last year that Strachan spent among the Anazzah Bedouin. This can be determined by the fact that at some point in that year he adopted a Muslim name as is required on conversion to Islam. The name he took was 'Mohammed Çelebi'. The choice is informative of his thinking at the time. The first name is the most revered among Muslims and must have pleased the emir, but Strachan viewed Mohammed as a false prophet, as is shown by comments he wrote in one of the Arabic manuscripts he bought at that time (Appendix: Strachan's Catalogue Ms. No. 19). Taking the name of the 'false' prophet to mark a 'false' conversion may have given him some private satisfaction. The second name 'Çelebi' was originally a Turkish title given to princes, but had evolved into a descriptor for a man of distinguished rank. So, even with his new identity, George Strachan was making it clear that he was a man of status and not a servant of the emir. Among the Arabs he became known as 'Doctor Mohammed' and began to be accepted as a full member of the tribe (Dellavida 1956: 49).

The next step in his transformation illustrates the extent to which his integration had progressed. The emir gave Strachan/Çelebi his brother's widow in marriage. This was a great honour for the Scotsman, but the action should be viewed as political rather than romantic. Marriage to the Bedouin princess meant that he became a member of the emir's family, not a servant or even simply a member of the tribe. Today in Muslim societies in the East, marriages of young women are usually arranged. In Strachan's time that was invariably the case. The girl, and often the groom too, had no say in the matter. However, this was not the case with widows, particularly in Bedouin society where personal freedom was more highly valued. Remarriage was not uncommon and although the wider family was involved in the proceedings, the woman had an important place in the decision making. If the emir's sister-in-law did not wish to live the rest of her life as a widow, she would have been able to initiate discussions on possible partners. The emir's permission would have been needed for any marriage to be concluded, but he would have had difficulty in forbidding it or continually turning down the procession of suitors, all of whom might have been politically problematic for him. Any Arab would have viewed marriage to the princess as a great prize. It would have placed them in a powerful position in the tribe, as they would have been able to challenge for leadership should the question of replacement of the emir arise. In other circumstances it has been known for Arab men to divorce their wives in order to marry into the ruling family. The princess, no matter what her personal attractions might have been, would not have been short of suitors. Feyyād may have been unwilling to add to possible rivals for his throne by allowing them to marry the widow, but while she remained unmarried the problem could not be ignored. His solution was to present the lady with George Strachan as her husband. Strachan could never challenge the emir for his position and, as well as removing the sister-in-law from the marriage market, their union bound the Scot more closely to the emir himself. Strachan's attitude to the marriage is unknown. At no time does Della Valle make any comment on his friend's feelings on the matter. The marriage undoubtedly took place and, although Strachan may have been less than wholehearted about it, the lady was enthusiastic if comments made by an official of the English East India Company are correct (Dellavida 1956: 48).

By 1617 Strachan/Çelebi had been almost fully absorbed into the life of the Bedouin. The question arises as to why Strachan stayed with the emir long after it must have been clear to him that he was in danger of being coerced into conversion to Islam. As a young man, he had sacrificed a great deal for his Catholicism. He had endured exile and been estranged from his family because he adhered to his faith rather than convert to Calvinism. It is difficult to believe that an agreeable lifestyle and a steady income would have been enough to make him pretend that he had made the much more dramatic change from Christianity to Islam. The reason that he remained and ran the considerable risk of his false conversion being discovered lies in his original purpose in travelling to the East. As he told Thomas Dempster, his decision was due to his wish to learn oriental languages, and he had succeeded in this to a remarkable degree. In a relatively short time living with the Bedouin, he had mastered Arabic to the extent that his speech was indistinguishable from that of a native speaker. Nevertheless, as a humanist scholar, his real interest lay in gaining a full and deep knowledge of the wealth of literature that Eastern societies possessed. Through the generosity of Emir Feyyād, he was able to start collecting high-quality texts many of which were unknown in the West. His intention was to return with them to Europe, where they would have constituted an asset which he could use to reinforce his reputation as a scholar. In order to safeguard this future, once he had studied the books, he began leaving them in the safekeeping of the Franciscan convent in Aleppo, depositing them on his visits to the city (Dellavida 1956: 66). Building up such a treasure took time and caused Strachan to stay with the Bedouins longer than was prudent, but there was an even more important reason for him to remain. He needed competent teachers to help him understand the significance of the texts and, indeed, even to make him aware of their existence.

From the outset of his journey to the East, George Strachan was short of money. If he had had unlimited funds, as was the case with Pietro Della Valle, he would have been able to indulge his curiosity regarding the sites of interest in the Holy Land and elsewhere, before settling down to study Eastern languages at one of the great centres of learning. This would have taken him to Cairo or possibly Damascus, since they were the principal rivals for academic supremacy in the Arab world where their literati vied for patronage from the Ottoman court in Constantinople (Elger 1984: 185–8). However, the learned elite of the imperial divan also recognised the excellence of scholars who lived in the eastern borderlands between their empire and that of Safavid Iran, in insignificant towns and villages such as Mardin and Amid. It was his extraordinary good fortune that, by becoming a respected member of Emir Feyyād's household, Strachan was able to gain privileged access to this remarkable group of scholars, renowned for their Islamic learning and knowledge of classical Arabic, Turkish and Persian literature. Through their tutelage he was able to gain an understanding of Eastern languages and literature that was far superior to that of any contemporary European scholar. His teachers were devout Sunni Muslim scholars who would not have engaged with a *kafir* in any other circumstance than that he was undergoing conversion to Islam and, more importantly, that he was being sponsored by the emir who was their own patron. If Strachan had not dissembled with them in his discussions on religious matters, they would have withdrawn their cooperation and he would not have benefited from their outstanding erudition. The existence of these eminent scholars in remote areas of Feyyād's desert kingdom, however, requires explanation.

CHAPTER SEVEN

The Ḥusaynābādī Scholiasts

Muslim Confessional Identities in Islamic Iran

In less than twenty years following the death of Mohammed in 632, Arab armies had conquered the eastern provinces of the Byzantine Empire and the whole of the Sassanian Persian Empire. The desert warriors had little experience of ruling such complex societies, and continued to employ the officials of the former administrations of their new territories to manage civil affairs and collect taxes. Initially in Persia the Pahlavi language and script continued to be used for this work, but in 697 their Arab overlords forced the introduction of *Kufi* script and, although the Persian language remained predominant, the elite of the administration became bilingual in Farsi and Arabic. Despite the fact that the Arab Empire was defined by Islam, it was tolerant of other faiths. In the continuing Arab wars of conquest, conversion of the conquered peoples to Islam was not compulsory but a significant number of non-Arabs did become Muslims. Identified as *Mawālīs,* initially these converts were second-class subjects especially in matters of taxation, but over time they gained equal status to Arabs. This was the experience in Persia and many of its elite became Muslim although the older religions such as Nestorian Christianity and Zoroastrianism continued to be followed. By the ninth century the Arab Empire had incorporated a large number of *Mawālīs* and the language as enshrined in the Qur'an was being corrupted. This problem needed to be addressed and it was a Persian scholar, Sibawayh, in Shiraz in 840 who first codified Arabic grammar and produced a dictionary of the language (Versteegh 1997: 4). By then, not only were the elite of the Persian administration bilingual but Persia had become a centre of learning for Arabic as well as Pahlavi literature. In the tenth and eleventh centuries, the Arab Empire in the East began to disintegrate, forming autonomous regions in Iran, Iraq, Syria and Anatolia. Native Persian rulers, and later Turkish invaders, took advantage of the Arabs' weakened position and gained control of Iran, extending their empire beyond Persia into Afghanistan and parts of northern India. This was the prevailing situation when the Mongols invaded in the thirteenth century capturing the greater part of the region before their advance was halted in Syria by the Mamluks of Egypt. Like the Arabs six centuries before, the Mongols used the existing administrations to collect their taxes. In Persia the educated elite, who carried out this work, were permitted to retain their culture ensuring that their literature was not only preserved but continued to develop. The invasion of the Turkic-speaking tribes led by Timur (Tamerlane) in the middle of the fourteenth century did not disrupt this tradition, and the Persian educated classes administered the empire for their Timurid overlords.

With each succeeding wave of conquerors, religious tolerance was maintained for Muslim and non-Muslim subjects. This was so even after 1294, when the Mongol ruler, Ghazan Ilkhan, converted to Shi'a Islam. The Arabs who had conquered Persia in the seventh century were Sunni Muslims. The schism in Islam between Sunni and Shi'a had occurred immediately after the death of Mohammed. His followers disagreed on the choice of the prophet's successor as Caliph (Leader of the Faithful). Some, who came to be known as Sunni, believed that Mohammed intended that his father-in-law, Abu Bakr, should succeed, while others (Shi'a) believed that it should be his son-in-law, Ali ibn Abi Talib. The two sides went to war over the question with each side revering their martyrs. The Sunni prevailed and became the larger grouping of Islam, but Shi'a adherents survived to develop their own traditions and liturgies, often as a persecuted minority. Although these sects constitute the major religious groupings within Islam, with Shi'a forming 10–15 per cent of Muslims, sub-divisions arose with the formation of mystical Sufi and Dervish *tarikats* (orders) which later were to play an important part in Persian society (Karamustafa 1994).

The erosion of religious tolerance in Persia started under the rule of the Safavid dynasty at the beginning of the sixteenth century. The Safavids were Oghuz Turks (Turkmen) who had developed their own religious identity under Shaykh Safi al-Din. At the beginning of the fourteenth century, Shaykh Safi established a Sufi order of Muslim mystics in Ardabil in Iranian Azerbaijan and his family became the hereditary grand masters of the order. In 1447, a century and a half after its foundation, one of Safi's descendants, Safavayga, converted the Sufi into a military order and espoused Twelver Shi'ism, the largest branch of Shi'a whose followers await the return of the twelfth imam, Muhammad al-Mahdi, whom they believe to be the spiritual successor to Mohammed. Safavayga's decision conflicted with what is the normal practice of Sufi mysticism, which does not restrict recruitment of its adherents to either Sunni or Shi'a. Inspired by their religious zeal, the Sufi Turkmen became a formidable fighting force attacking Christian Georgia and dominating much of the Caucasus and western Iran. One of Safavayga's descendants, Ismail, on becoming grand master of the order, claimed that he was al-Mahdi (Küçükhüseyin 2017: 229–39). Believing in his invincibility, Ismail's followers achieved stunning military successes, capturing much of what had been the Timurid Empire. In 1501 he declared himself 'Shah Ismail of Persia' and decreed that Twelver Shi'ism was the official religion of his empire and all other forms of Islam were heretical. To enforce this decree, he began to persecute Sunni Muslims. On the capture of Baghdad in 1508, not only was the Sunni population slaughtered but their shrines and tombs were desecrated. The following year he turned eastwards and conquered Afghanistan but that marked the high point of his success and the maximum extent of his empire.

Ismail had recruited his army in part from among the Dervish Çelali tribesmen of eastern Anatolia. They became known as Qizilbash (*Qezelbāš* or Red Hats) from the colour of the pointed caps in the middle of their turbans, which they wore as a sign of loyalty to their grand master. They were fellow Oghuz Turks but their lands were under the suzerainty of the Ottomans. When they joined with the Safavids, they rose in rebellion against their Ottoman overlord, Sultan Selim I, 'the Grim', who declared them renegades and heretics (Zarinebaf 2011: 1–35). He crushed the rebellion but considered that the new shah was a military threat that needed eradicating (Zarinebaf 2011: 1–35). In 1514 his army decisively defeated the Turkmen and occupied Baghdad as well as the Safavid territories of Kurdistan and Armenia. Once defeated, Shah Ismail lost his aura of invincibility as al-Mahdi, and many of his Turkmen changed their allegiance to the Ottomans. Two years later Sultan Selim defeated the Mamluks and extended Ottoman power over Egypt as well as Syria and Iraq. As the undisputed Commander of the Faithful and Guardian of the holy sites of Mecca, Medina and Jerusalem, he decreed that Sunni Islam was the official religion of his greatly enlarged empire. Although weakened, the Safavids persisted in threatening Ottoman power through continuing support of Qizilbash insurrections and giving shelter to Ottoman princes who were vying for control of the Sublime Porte. Confrontations between the two empires continued under successive rulers. Sultan Suleiman I, 'the Magnificent', carried out major military campaigns in 1535 and 1545 against Ismail's successor, Shah Tahmasp I. He conquered more of the Safavid territory in the west and brought order to the Ottoman Empire's eastern borders, but he did not complete the conquest of Persia. There was no serious attempt made to do this. Suleiman concentrated his major military efforts on extending his empire in Europe. Shah Tahmasp was able to retain his much reduced empire, but he had lost not only the Caucasus but his family's homelands in western Iran including his father's capital, Tabriz. By the middle of the sixteenth century, the borderlands between the two empires had stabilised, with the Ottomans being the dominant power, but the two empires had defined themselves as enemies in religious as well as political terms.

Custodians of a Great Heritage

It fell to Tahmasp's grandson, Shah Abbas I, 'the Great', to regain control of all of his great-grandfather Ismail's empire. George Strachan was to spend many years in Abbas' Persia and was witness to much of what the shah had achieved, but part of Abbas' strategy lay in reducing the dominance that Qizilbash tribesmen and the old Persian nobility had over the empire and the shah. They held all the powerful positions in the administration of government and, most importantly, the military (Chapter 9). Among other actions Abbas decreed that state administrators and officials espouse Shi'a Islam; those who did not were forced to flee Iran. These exiles included a number of the elite who were the custodians of Persia's rich heritage of literature. These scholars went westwards into the lands of the Sunni Ottomans, taking with them their libraries of Persian and Arabic texts some of which predated the Mongol invasion of three and a half centuries earlier. They were city dwellers but, in the hope that they would be able to return to Persia, they chose to settle as close to their homeland as possible.

The scholars set up their new schools in towns and villages in Kurdistan and along the fringes of the Great Syrian Desert. Due to Shah Abbas' continuing wars against the Ottomans, it was not possible for them to create permanent establishments in the borderlands, and they moved from time to time, taking their students with them. These schools were not typical of the rustic madrassas to be found elsewhere in the Islamic world. Their renown was as great as that of the educational institutions to be found in Damascus and Cairo and was based firmly on the erudition of their leading scholars. As well as being custodians of their ancient heritage, they were scholiasts who wrote profound interpretations on classical Arabic and Persian works. In addition the scholars were noted for the dynamic method of teaching religion that they had developed. They did not restrict their studies to the Qur'an but drew on a wide range of philosophical, philological as well as theological texts in classical Arabic and Persian. Their concern was not solely with the content of the chosen books, although they were acknowledged masters in that; in addition they made great efforts to reform the didactic practices of learning by rote used at the time, to ensure that students gained as profound an understanding as possible of the true meanings of what were ancient and often obscure texts (Endress 2006: 371–422). The situation in Iran never allowed them to return to their ancestral homes, and over generations their families remained and flourished in the borderlands of the two empires.

Early in his reign, Emir Feyyād had given refuge to Sunni exiles from Shia Iran including some of these Persian scholars. The emir protected the exiles as an act of charity to fellow Sunnis, but their presence in his domain enhanced his reputation not only as a benevolent ruler but as a patron of learning. Among the most prominent of the scholarly exiles was the family of Ḥusaynābādī (Klein 2007: 139). Detailed knowledge of these scholiasts exists only from the 1610s, the time of Strachan's sojourn in the desert, and is based in large part on the writings of Sari Osman, a graduate of a madrassa in Constantinople, who in the middle of the seventeenth century travelled east to study with what he described as 'the renowned scholars of the borderlands', the most prominent of whom was Ahmad Haydarani of the Husaynābādī family. Ahmad's grandfather, Muhammad, had fled north-west Iran when Shah Abbas began his persecutions, and his grandson had been born in Kurdistan. From his youth, Ahmad, along with other members of his family, was involved in studying and later teaching in the family madrassa. Sari Osman spent time as a student of Haydarani who by the time of his visit had had a long and distinguished career as a scholiast and teacher (Swartz 2017: 1–34).

According to Osman, Haydarani corresponded with scholars in Constantinople and Damascus and the scholia he wrote on difficult texts were respected and highly sought after. One of his early correspondents was a prominent Turkish bibliophile, Katib Çelebi, resident in the Ottoman capital. It may be mere coincidence that Strachan called himself Çelebi when he 'converted' to Islam since the name, although not common, occurs among Turkish families. However, it is possible that the Turkish scholar was known to Strachan through his conversations with his teachers. Although Ahmad was still a young man in his twenties when Strachan was a member of Emir Feyyād's divan this would not have precluded his involvement along with his father and possibly grandfather in Strachan's education in Islamic studies. Haydarani's scholarly tradition was continued by his son, Haydar, and his grandsons all of whom, like Ahmad, were well known beyond the borderlands. The Husaynābādī family remained active as scholars running their madrassa until the end of the eighteenth century, all of which testifies to the intellectual rigour of the scholars of the borderlands (Swartz 2017: 3–6).

None of the written sources on George Strachan mentions the names of the tutors he met in the desert but Emir Feyyād ensured that they were the best available. If the Husaynābādīs were not involved, the scholars Feyyād chose would have been of similar background and quality. Pietro Della Valle makes this clear in his journal when he describes Strachan's discussions with these intellectuals:

> [Strachan] continually elicited discourses by the most accomplished literati among the Arabs, whom the Emir gathered about Strachan with a view to his conversion, to say nothing of his obtaining the perusal of any book that he desired, which the Emir either furnished him with or obtained for him. (Della Valle 1664: vol. 2, 719–20; Yule 1888: 317)

They could not refuse the emir's request to accept the *kafir* as a pupil, and Strachan could not have wished for better mentors in his study of Eastern literature. By accepting the tutelage of teachers of the status of the Ḥusaynābādī, he was embarking on a course of higher study which he must have found as demanding and rewarding as any he had encountered at universities in Europe. These circumstances in large part explain why Strachan stayed with Emir Feyyād longer than was prudent. He would have ended this advantageous arrangement only with extreme reluctance. According to Della Valle, it was through his discourses with these teachers that Strachan 'gained an excellent knowledge of the Arab language, as well as the fullest acquaintance with the most abstruse matters of the Mahommedans' and "every day becoming more and more master fundamentally of the most intimate details of Mohammedanism"' (Della Valle 1664: vol. 2, 719; Yule 1888: 317).

The instruction that he was receiving and the conversations that he engaged in with his Islamic teachers involved reading a great number of texts. Through his teachers he had access to a considerable library of works in Arabic, Farsi and Turkish (Dellavida 1956: 47). But these books were not available for purchase.

Buying books was only possible in the large cities. The literature of the Ottoman Empire of the time had yet to undergo the revolution that printing technology had caused in the West, and in order to own a book it was normally necessary for scholars to have it hand- copied. Strachan's library shows that his surviving collection consists of copies that he had commissioned and others he was able to buy second hand. The professional scribes whom he used lived in Aleppo and Baghdad, but he may also have engaged students at the desert madrassas to copy texts owned by his teachers. Some books in his collection contain the scribes' entries, making it clear that they had been produced in Baghdad. Others make no mention of the scribe or where he was working which suggests that it was not considered significant, as would have been the case in a village madrassa. Strachan has added his own comments in Latin on the flyleaf giving the book's title and date of purchase, stating whether or not he had bought it or had it copied and a brief description of its contents. From these entries it is clear that he had access to Arabic and Persian texts other than through his Muslim tutors, and that his dealings with them were in part on his own terms. He requested and received books which covered the subjects that interested him, and his challenge to the scholars was to 'convince him that their religion was better than his own' (Della Valle 1664: vol. 2, 719; Yule 1888: 317).

Intellectual Debate

Strachan's early education had been largely at the hands of the Jesuits who used a highly structured approach to teaching. He had been taught methods of rational disputation that were laid out in the *Ratio Studiorum.* These lessons would have stayed with him throughout his adult life, and when he was involved in any discourse it would have been natural for him to adopt the precepts of 'Thesis: Antithesis: Synthesis' in which he had to answer his opponent's argument (thesis) with a counter-argument (antithesis) but then attempt to find common ground between the two positions (synthesis) in the search for an acceptable reconciliation of differing points of view (Müller 1996: 343–5). In Jesuit training, such debates were conducted as part of a procedure known as *concertatio* in which the protagonists engaged in 'honourable rivalry' in the presence of their classmates and professors (Fitzpatrick 1933: 203). In his discussions with the

Muslim scholars at the madrassas, Strachan would have relied on his early Jesuit training to counter their arguments without appearing to be hostile to their ideas. His privileged position with the emir and their continued cooperation in his education in religion and literature depended on their believing that he was open to their arguments.

Their discussions must have lasted for the greater part of the two years that Strachan spent in the desert, and would have centred on debating the 'Five Pillars of Islam' which constitute the basic tenets of the Muslim religion. The first and most important of the 'pillars' is that 'there is only one god and Mohammed is his prophet'. Acceptance of this assertion is sufficient to constitute conversion to Islam. Using casuistry in a manner for which the Jesuits gained a reputation among their enemies, Strachan appears to have explored the points on which the two religions agreed without deviating from his own beliefs. He seems to have thought that, by engaging in discussion in this manner, he had avoided expressing acceptance of both aspects of the vital first tenet. Nevertheless, his hosts believed that he had accepted them. Discussions around the second, third and fourth 'pillars' would have presented him with little difficulty. 'Regular prayers, charitable giving and fasting (during Ramadan or Lent)' were virtues strongly encouraged by both Islam and Christianity. The fifth and last 'pillar' requires that good Muslims 'visit Mecca once in their lifetime if possible'. Although it would have had no doctrinal significance for him, the wandering Scot would not have seen this as burdensome but rather would have looked forward with interest to such a journey. As matters turned out, Strachan had no opportunity of visiting the Muslim holy city. At the end of his discussions with the Islamic scholars he believed, as he told Della Valle, that he had not renounced his Christianity (Della Valle 1664: vol. 2, 720). Nevertheless, he must have altered his Christian liturgical practices of prayer and fasting to conform to those of his Muslim hosts. Strachan would have been able to rationalise this behavior. The five daily calls to prayer for Muslims could be considered analogous to the call to prayers at the canonical hours of the *Horarium* which was followed by Christian regular orders of monks and nuns. He was completely familiar with these practices and had joined in their communal prayer services during the many occasions when he lodged as a guest at religious houses. Likewise, observing Ramadan would not have conflicted with the continuation of private fasting during Lent. Although Strachan in his own conscience remained a Christian, by conducting his discussions in the manner he had and altering his outward behaviour to accommodate Islamic liturgies, he led his tutors to believe that their arguments had persuaded him that Islam was the true faith.

Flight

Strachan's interest in these discussions was intellectual and he had not contemplated changing his religion. His justification for his actions was to gain a deeper understanding of Eastern languages, culture and literature which inevitably involved studying their holy texts. This would have been impossible without the help of his tutors. They would not have cooperated with him under any circumstances, other than that he was prepared to convert to Islam. It was forbidden for unbelievers even to possess a copy of the Qur'an, and serious theological discussions with a *kafir* would have been anathema to them as devout Muslims. Della Valle later wrote in his journal that by engaging his teachers in such debates, Strachan was enlightened on the 'most abstruse matters' (Della Valle 1664: vol. 2, 719–20; Yule 1888: 317). His teachers supported their arguments with appropriate quotations from a wide range of sources: classical Arabic and Persian writings and theological texts. They did so to further their arguments, but these references would have provided Strachan with unique guidance as to which books he should try to include in his own collection.

Initially Strachan's time spent in the desert with the Bedouin must have seemed idyllic. As Della Valle wrote, 'in this life he found, by his own account, the utmost relish' (Della Valle 1664: vol. 2, 720; Yule 1888: 317–18). He was furthering his ambition of learning Eastern languages and literature under the tutelage of distinguished scholars and accumulating a library of valuable texts. His hosts not only respected his erudition but understood that he was a gentleman by birth and treated him accordingly. He was being paid enough money to indulge his interests as well as save for future needs. Complete satisfaction with his life was impeded only by the insuperable obstacle of the emir's requirement that he convert to Islam. He had shown his devotion to Catholicism throughout his life and, judging from glosses that he added to some of his manuscripts, he appears to have had an intellectual contempt for Islam which closer examination had intensified (Appendix: Strachan's Catalogue Ms. No. 34).

When he joined the emir's household in 1616, he expressed an open mind on the merits of Islam rather than risk his new position by giving an outright rejection of his employer's overtures on conversion. By doing so he benefited not only from receiving a comfortable living but more importantly from being introduced to the eminent scholars, who were prepared to share their wealth of learning on the condition that he was actively considering conversion to Islam. Strachan kept presenting arguments to them which they refuted according to Islamic tenets, but by late 1617 either he ran out of excuses for prevarication or the emir's patience was exhausted. The timing of his acquiescence can be deduced from the fact that, with one exception, all of the books in his collection that date from his time in the desert have inscriptions referring to him using his Christian name, *Georgius Strachanus Merniensis Scotus.* Only one is known that uses his Islamic name stating that it was copied 'on behalf of the Frank physician Mohammed Çelebi' (Appendix: Strachan Catalogue Ms. No. 36). The book is undated, but can refer only to the period in the desert after Strachan's 'conversion'. Given that he was acquiring books frequently, at the rate of no fewer than one every two months, this would indicate that there would have been a short interval between the 'conversion' and his leaving the emir's household.

During this period a number of events happened to Strachan in quick succession. After his Islamic tutors had affirmed that his beliefs were fully compliant with the teachings of Islam and that he was a true Muslim, Emir Feyyād arranged the marriage between his sister-in-law and the Scot. Probably at the insistence of Strachan's new wife, the emir made it clear that as a Mohammedan husband he must be circumcised. Circumcision is not a religious requirement but it is an almost universal custom among Semitic societies. For 'cradle' Muslims and Jews the operation is carried out while very young and normally has no deleterious effects. Circumcision of adult males is riskier due to the longer period needed to heal and the possibility of infection. This aspect of the request may have weighed on Strachan's mind, but it is more likely that he baulked at the procedure being a physical symbol of his 'conversion'. Once he had subjected himself to circumcision, it would have been impossible for him to maintain the argument that he had not converted to Islam. He was determined that this should not happen, but Feyyād was insistent. Strachan had to try to escape. Della Valle wrote of what his friend later told him of this time in his life:

> But, at last, seeing that the Emir was becoming more and more stringent in pressing him to undergo circumcision, he could no longer put off his retreat. And finding an opportunity when the Emir's camp was in a certain tract not far from Bagdad, he successfully arranged his escape, not without a good deal of trouble and disturbance on the part of the lady who deemed herself his wife, and got away into that city, where he stayed several months, during which the Arabs never quite lost the hope of getting him back. (Della Valle 1664: vol. 2, 720; Yule 1888: 318)

Even making full allowance for all the attractions of his life at the emir's court, it is clear from Della Valle's account that Strachan had stayed longer than he should. As an intelligent man, he would have been able to foresee much of what befell him and should have acted sooner to avoid it. Apart from his reluctance to give up what he saw as a pleasant life, he had to delay his departure until it was safe. Strachan was to some extent a prisoner and, while the Anazzah Bedouin migrated across the desert taking their herds to new pastures, he had no opportunity to leave without being detected and restrained. Trapped in this situation, he had no alternative but to agree to the emir's wishes on conversion and marriage. He may not have viewed his deception regarding his 'conversion' as serious and he may even have been willing to continue in the emir's service with a new wife, but the matter of circumcision had made his position untenable. He decided he had to leave his Bedouin life behind him.

His plan for leaving required that the Bedouin encampment was within easy travelling distance of Aleppo or Baghdad, both of which were outside the emir's jurisdiction. In late 1617 or, more credibly, in early 1618 when the Bedouin camped near Baghdad, he took his chance. At some point in the previous year he had commissioned a scribe in Baghdad, Ibrāhīm ibn al-Ḥasan, to copy a prayer book, *aṣ-Ṣaḥīfa al-kāmila,* for him (Appendix: Strachan's Catalogue Ms. No. 43). The scribe completed the work in February 1618, and news from al-Ḥasan that the book was ready for collection may have been the excuse that Strachan needed to visit Baghdad. Both the date of completion and Strachan's note regarding the time and place of purchase are written on the fly-leaf and are persuasive of Strachan having left Emir Feyyād's camp in early 1618. If he had left later, al-Ḥasan would have arranged for the delivery of the book to him at the desert camp in order to receive payment. His flight was against the wishes of his wife who made her views known vociferously but the Bedouin believed that he was taking another one of his occasional trips into Baghdad on personal business. In such a male-dominated society, it was not a wife's place to determine her husband's actions and the tribesmen would not have interfered in such a dispute. While travelling in the desert, he had become adept at packing his personal belongings and loading pack animals, ready to move on at short notice. Strachan escaped on horseback taking his books and savings. Dressed in his Arab robes he arrived in Baghdad free of the Bedouin (Dellavida 1956: 50). Strachan behaved as he had done on previous visits, but on this occasion he remained for more than a year reading in the city's libraries and buying a great many more books to add to his collection. Judging by the translations and textual glosses that he made, he spent his time in detailed study of his purchases and improving his understanding of classical Arabic and Persian literature. It is through the work that Strachan carried out on his library that he made his greatest contribution to European understanding of Arabic literature and Islamic studies. This library is the reason that Strachan deserves to be recognised as more than a footnote in the history of others and it requires examination in its own right.

CHAPTER EIGHT

Strachan's Library

Public Libraries

Until the invention of the printing press in Europe, the Arab Islamic civilisation was arguably the most literate and bookish society in the world. From as early as the ninth century, Middle Eastern scholars had benefited from a revolution in book production, driven in large part by the introduction of paper. While Europe was using prepared animal skins to make books, Arabs had gained the secrets of paper production from the Chinese. Chinese paper was hand-made. The Arabs greatly improved on the technology by harnessing watermills to operate trip hammers which pounded rags to make the long-fibred pulp needed for the finished paper. In this way they mechanised the most labour-intensive part of papermaking. Arab linen paper was much cheaper to produce and of more consistent quality than parchment or vellum. Papermaking was carried out throughout the Arab Empire from Central Asia to Spain, with Baghdad being the centre of greatest production.

Inexpensive paper allowed books to be produced more cheaply, which led to the foundation of libraries on a scale unknown in Christian Europe where paper technology was not used until the thirteenth century. By that time books were widely available throughout the Islamic world. The great cities of the Middle East developed a tradition of endowed local libraries. Madrassas, mosques, teaching institutions and even the mausoleums of prominent people possessed libraries available for the use of the literate public. As well as public libraries, wealthy individuals had private collections which surpassed almost any available in Europe. Although there are no quantitative records, recent scholarship has arrived at estimates of literacy in excess of 10 per cent among city-dwelling Arabs (Gründler 2016: 31–66). An analysis of the catalogue of an Arabic library of the thirteenth century, that of the Ashrafiya madrassa in Damascus, has shown that this undistinguished establishment had almost 2,000 book titles. The works were on theology, poetry, mathematics, the sciences and history written by ancient Roman and Greek authors as well as great Islamic scholars. The Christian West had little to compare with this modest library of the Arab world. The book collections of all the colleges of Cambridge University combined did not exceed that of the Ashrafiya until two centuries later (Hirschler 2016: 2–3).

When George Strachan visited Aleppo and Baghdad, books were still more generally available there than in Europe even though the Ottoman Empire had not adopted the printing press on a widespread basis. As personal physician to Emir Feyyād, he had both authority and money which gave him access to the cities' public libraries as well as to the private collections of booksellers. In the course of the year he spent in Baghdad after leaving Feyyād, he took full advantage of this privilege. Judging by the catalogue of the Ashrafiya madrassa, there would have been thousands of texts available for him to examine. Before deciding on which commissions to give to the scribes, he would have been able to indulge himself by searching through books on a wide range of subjects by many authors. In the short time he had available during his earlier visits, it would have been impossible for him to read more than a small portion of this material which would explain why he tarried so long in the cities before having to return to the emir's encampment.

Cost of Manuscripts

Although a great deal of information survives regarding the books he collected, it is far from complete. In 1622, while staying at the monastery of the Discalced Carmelite Friars in Isfahan, he entrusted a large part of his collection to Fr Vincent of Saint Francis who was returning to Rome (Chapter 11). The records of the order show that Fr Vincent received sixty-one volumes in Arabic, Persian and Turkish and was given authorisation by Strachan to retrieve an unspecified number of books that he had left for safekeeping in Aleppo. Strachan had numbered his volumes and included a catalogue for Fr Vincent's benefit but the catalogue has not survived (Chick 1939: 235). At least twenty of the books are missing. Forty have been identified in the Vatican Library and the National Library in Naples (Dellavida 1956: 73–5). Others have been identified in Cambridge University Library (Browne 1896: 46), the British Museum (Dellavida 1956: 108) and the Majlis Shūrā Library in Tehran (De Young 2015: 132).

Most of these were acquired by Strachan after he had left Emir Feyyād, but a dozen contain dated inscriptions showing that he possessed them during his time in the desert. There are none that predate his employment with the emir. His finances prior to that did not allow him the luxury of such purchases. While resident in Baghdad in 1618, Strachan bought a minimum of twenty-one books. It is from these purchases that we know that he was conversant with Persian as well as Arabic. Many of the people who lived in the borderlands between the Ottoman and Safavid empires were bilingual including Strachan's Islamic teachers. Three of his Persian manuscripts remain in European libraries but none of the Turkish texts, which he is known to have possessed, have been discovered. Strachan did not collect books simply for the pleasure of acquisition. He spent time studying the texts in detail. His intention was to advance his understanding of both language and context in relation to the cultures that produced them. In addition it was important to him that he accrued an educational resource which, on his return to Europe, he could use for research, teaching and publication. For this reason his library needed to be as large and eclectic as possible.

Collecting books became a priority for Strachan, with the only constraint appearing to have been the depth of his pockets. The cost of the complete library is impossible to ascertain but would have been considerable. In 1624 in Isfahan he bought an Arabic copy of Euclid's *Elements* (Appendix uncatalogued Ms. 4 of 6) for 15 abbasi – little more than one pound sterling. The text had been copied in 1455 and, although the scribe had taken care in his handwriting and with the geometric diagrams, it was full of omissions and errors. The scribe did not appear to have been familiar with mathematical notation, since he did not correct even simple errors. As a text it is superficially presentable, but can best be described as unfinished (De Young 2015: 139–40). In this condition, it could not have commanded more than a modest price. Other examples in Strachan's surviving library are of much higher quality, and must have been more expensive.

It is difficult to compare the cost of such books with European printed texts. Because of the labour involved, handwritten books were more expensive than equivalent printed texts. European booksellers made conscious efforts to keep prices low to encourage sales and it was normal for them to sell the texts unbound. Buyers could then decide how much they wanted to spend on the bindings. The wealthy commissioned expensive leather covers embossed with the family crest, while impecunious scholars used them unbound (Ammannati 2018: 161–77). Bindings for manuscript copies in the East were expensive also. It was common practice for several books to be bound together to save on expense. Judging from the seven books in his surviving library that still have their original bindings, all of Strachan's purchases appear to have been expensively bound. Two were bought second-hand while the other five were copied for him. In each case the cover is red or black leather and two of his commissioned copies are embossed with gold leaf. Strachan did not stint on these status-enhancing finishes, which would have made the volumes more expensive.

His purchase of the flawed copy of Euclid illustrates another aspect of Strachan's attitude to acquiring books. When he could, he rebelled against the frugal life he was forced to endure by indulging himself in a book purchase. He had shown this trait as a young man when he bought his album amicorum which was leather bound with his coat of arms embossed on it in gold leaf. The acquisition of something of value representing his education and social status, when he had restricted finances and limited prospects, appears to have helped him maintain his dignity as a gentleman.

During the initial part of his stay in Iran, Strachan was employed by the English East India Company on an annual salary of £40. This was increased for a few months to £100 before he was dismissed by the company. In 1624, when he bought the *Elements*, his employment had been terminated

(Chapters 9 and 10). The expenditure he made on the books he collected while living in the desert with Emir Feyyād exceeded his entire income while working for the English merchant company. He had met the expense by using up almost all of the savings he had accrued during his service with the emir. His ability to save this considerable sum at the same time as equipping himself in the style of a Bedouin prince with the display of wealth that this new status required proves that Emir Feyyād was a liberal employer who rewarded the Scotsman generously.

Strachan's Interests

After he had exhausted his savings, Strachan could buy books only occasionally as his reduced finances allowed. Nevertheless, his library was extensive and covered a wide range of subjects. The collection gives an insight into his interests and hints at the reasoning behind his choice of texts. Nine books survive dealing with Arabic grammar, rhetoric and logic – the three core elements of the Trivium which, in Latin and Greek, formed the basis of his early education in Jesuit colleges. These books would have provided Strachan with materials essential for the teaching of Arabic on his return to Europe. The missing element in this collection is an Arab dictionary although it is possible that it was among the volumes that have been lost. He included additional books which he considered to be 'very useful to students of Arabic' (*Liber Arabicantibus utilissimus*) (Appendix: Strachan's Catalogue Ms. No 54), a clear indication that he was equipping himself to teach oriental languages on his return to Europe. There are no records of Persian or Turkish texts of this nature which might indicate that he did not intend teaching these languages.

Most of the Persian books he collected were on history and literature and appear to have been for his own interest and use. None of the Turkish texts, which are known to have existed, have survived and may have been, like the Persian texts, for personal enjoyment. Four mathematical and astronomical texts from his library remain, and these include the first book written in Persian that we know he possessed. This was a textbook on mathematics and astronomy, *Risāla-I Mu'īniyya,* which summarises some of the work of the famous fourteenth-century Persian mathematician and astronomer, Naṣīraddīn Ṭūsī (known in Europe as Tusi) (Appendix: Strachan's Catalogue Ms. No. 59). Strachan's earlier life as a teacher of mathematics in Paris may have inspired his choice, and he may also have considered them to be of use on his return to Europe. Professor Dellavida expressed surprise that Strachan did not collect books on medicine at a time when 'Arabic medicine was highly appreciated even in the West'. He assumed that Strachan 'cared little for the science' (Dellavida 1956: 102). He based his remarks on the fact that no such texts have survived. A more likely explanation of their absence is that Strachan did not place them in the care of Fr Vincent, but retained them for his own use. Although no longer formally employed by the East India Company as a physician, his skills would still have been in demand. On taking up his appointment with Emir Feyyād, he obtained some medical books from his Flemish friend in Aleppo. It would have been in keeping with the Scot's nature to add to this store with Eastern texts (Chapter 5).

In addition to books which had a practical purpose, he made other purchases seemingly merely to satisfy his natural curiosity. They range from the esoteric matters of magic and divination of dreams (Appendix: Strachan's Catalogue Mss Nos 55, 61) to two apocryphal texts that purport to have been written by Aristotle: one a pamphlet on the tactics of warfare imagined as having been written as advice to his pupil Alexander the Great (Appendix: Strachan's Catalogue Ms. No. 35). This would not have been expensive but, by his own account, he spent lavishly on a fifteenth-century copy of *Ḥalbet al-kumyat* by an-Nawājī, a popular set of stories on the joys of drinking and other pleasures forbidden under Islamic law (Appendix: Strachan's Catalogue Ms. No. 15). The cost was due to its being a finely decorated calligraphic copy. Strachan described it as rare but, despite or perhaps because of its prohibited subject matter, cheaper copies were relatively common. These subjects sit so far outside the main themes in his library that the impression is given that they were bought on a self-indulgent whim rather than to a fixed plan, but they make up only a small portion of the book collection.

The bulk of Strachan's library consisted of volumes dealing with Arabic and Persian literature and Islamic studies. These were to be of great interest and immense value to Western scholars, and it is on Strachan's acquisition and interpretation of them that his real service to European scholarship rests. When he began his life in the desert, his first need was to master the Arabic language in order to debate with his teachers. To help in this, he sourced books on grammar and Qur'anic and Islamic studies but, even at this early stage of his exploration of the language and culture, he began to collect books of poetry, initially in Arabic but later also in Persian. It is clear from the number of books bought that he had developed a true appreciation of the work of Eastern poets. Twelve volumes of poetry survive, together with a further six books of *belles-lettres* that deal with proverbs and other subjects that are poetic in nature. His interest in poetry seems to have lasted throughout his stay in the East (Appendix: Strachan's Catalogue Mss Nos 13, 18; also Ms. No. 19 in Persian). One of his most expensive purchases was a collection of poetry by Abdaljabbār ibn Abī Bekr, a twelfth-century Arab scholar living in Sicily. The manuscript was completed in July 1210 and is one of only three such copies known to exist (Appendix: Strachan's Catalogue Ms. No. 29). It is the oldest and most precious text in Strachan's surviving collection. Judging from the style of script used, the book's scribe lived in Spain. Inscriptions on the fly-leaf show that in the late fourteenth century it was owned by a famous Egyptian scholar, Abdallāh ibn al-Ḥasan al-Awḥadī, before being bought forty years later by a resident of Baghdad.

Arabic Poetry in Europe

Access to such treasures put Strachan in a unique position. As a European humanist he had written a number of poems in Latin, some of which had met with genuine critical approval, and when presented with the opportunity to explore the works of Eastern poets, which were abundantly available in public libraries, it is not surprising that he indulged his interest to the full. Such privilege was beyond the reach of his contemporaries in Europe, some of whom had begun to appreciate the genre. In his *Oration on the Value of the Arabic Language* (1620) Thomas Erpenius extolled the virtue of the work of Arab poets. He urged his fellow scholars to study these poems for their 'elegance of invention, as well as learning, care in composition, and sweetness in harmony and rhythm' (Loop 2017: 230–1). Erpenius' short-lived appointment as professor of Arabic at the University of Leiden from 1620 to 1624 marked a major turning point in the European study of the language. Prior to Erpenius, Christian Europe's motivation for studying Arabic stemmed from the desire to mount theological attacks on Islam. Although this was to remain a major reason, the Dutch scholar's intervention began the appreciation among northern European scholars of the literature of the East in its own right. Erpenius published an Arabic grammar (*Rudimenta linguae Arabicae,* 1620) specifically to encourage others to take up serious study of Islamic literature, especially poetry.

Initially his followers had only limited access to samples of Arabic poetry through earlier publications of the Qur'an which had been inadequately translated and a few works such as the series of funeral orations included by Leo Africanus in his *Description of Africa* (1600). Despite growing interest, scholars were constrained by the paucity of accessible texts. Equally important, progress in understanding was hampered by the almost complete absence of native Arabic-speaking tutors able to explain the subtleties of the great works. Jacobus Golius, Erpenius' pupil and successor as professor at Leiden, published the first printed copy of poetry in Arabic script but it was 1629 before he achieved this (Loop 2017: 233). Strachan had left Europe nearly a decade before Erpenius was appointed to his post in Leiden. During his time in Baghdad and Persia, Strachan had gained a level of both understanding and appreciation of classical Arab and Persian poetry that was well in advance of his contemporaries in Europe.

When Fr Vincent returned to Rome with Strachan's library in 1622, the poetry it contained should have been of immediate value to scholars. However, the volumes appear to have been left largely undisturbed for a number of years. They had been entrusted to the Carmelites' safekeeping awaiting Strachan's return from the East. He had promised that on reclaiming the books he would pay for their transport and storage (Chapter 11). He never returned but, during the period when it was possible that he might, the friars do not appear to have made the existence of these texts generally known. Strachan's poetry library would have been a major advance on what was available to European scholars at the time. The quantity of material and, more importantly, the instruction on its interpretation that Strachan had provided by means of his translations and glosses would have been revelatory. Two decades after Strachan's library had been lodged in Rome, Golius and Guadagnoli were still struggling with the basics of works in Arabic. Filipo Guadagnoli published his Arabic grammar, *Breves arabicae linguae institutiones* in Rome in 1642. In it he used Arabic poetry to illustrate grammatical points, but he was forced to be highly selective of the verses chosen since, as he repeatedly wrote, he could not understand certain passages (Loop 2017: 236–7). It is clear that if Strachan had returned to Europe along with his books, his services as a teacher of Arabic would have been in great demand. With the wealth of literature, especially poetry, that he had at his disposal, he would have been acclaimed as one of the most distinguished orientalists of his time. In addition, great works of Arab poetry would have become known in Europe much earlier than many of them did.

Strachan's Proficiency in Arabic

Such a claim is justifiable because Strachan had a mastery of language and literature that was well beyond the abilities of Erpenius and Guadagnoli or any of their followers. No European scholar came close to his level of understanding until late in the seventeenth century. Proof of this assertion has been provided by Professor Dellavida. While researching Strachan's life, Dellavida examined the surviving books of Strachan's Arabic collection. Using the synopses of their contents, the glosses and inter-linear translations that Strachan had provided, he was able to gauge the degree of understanding that Strachan possessed. He did not carry out an assessment of his interpretation of Persian texts and, since none of the Turkish texts have survived, no assessment of Turkish is possible. At the time he wrote Strachan's biography, Professor Dellavida was well placed to carry out an assessment of the Scotsman's proficiency in Arabic. He had held professorships of Arabic and Semitic Languages in universities in Naples, Turin and Rome as well as being professor of Arabic in the University of Pennsylvania. He sourced thirty-seven volumes collected by Strachan, twenty-five of which are held in the Vatican Library and twelve in the National Library in Naples. Taking account of the dates when Strachan obtained these books, Dellavida was able to assess the progress over time that the Scot had made in his appreciation of the subtleties which the texts presented.

The earliest surviving book was a purchase made by Strachan in Aleppo in 1615. Dellavida was able to deduce from the glosses that Strachan had a good understanding of its contents which were on Arab grammar and rhetoric. However, he noted that in the book's title Strachan had translated the Arabic *talkīs* as 'explanation' and not 'summary' as it should be (Appendix: uncatalogued Ms. No. 1 of 6). In his acquisitions of 1616–17, the Scot had made a few similar errors which, although not a fully accurate rendition of the texts, did not distort their meaning (Appendix: Strachan's Catalogue Ms. No. 37). These were the books that Strachan had collected during his service with Emir Feyyād in the desert. There were no such errors in his translations of texts that he purchased after that time although Dellavida did identify a recurring misinterpretation of a different nature. Two of the books he possessed while in the desert Strachan described as 'poems in praise of the accursed Mahomet' (*poemata in lau[dem] Mahometis Maledicti*) (Appendix: Strachan's Catalogue Mss Nos 34, 36). Dellavida takes pains to point out that these poems contain *ḥadīths* and that Strachan had failed to appreciate their significance as acts and sayings attributed to the Prophet. His Muslim teachers could not have explained this to him. The reason for their apparent lapse may have been that they were reluctant to use the actual words of the Prophet in their discussions with a *kafir*. After his flight from Feyyād, when he was pursuing his own studies in Baghdad in 1618, Strachan finally recognised the true meaning of *ḥadīths*, a fact that he made clear in his notes in *Ma'ānī al-āthār*, one of the books he purchased at that time (Appendix: Strachan's Catalogue Ms. No. 52). The last failing that Dellavida identified in Strachan's understanding was shown in his synopsis of the expensive book of racy short stories and poems that he bought in 1618. Because of the nature of its contents, it is unlikely that he discussed it with a native Arabic-speaking Muslim and perhaps due to this he failed to understand the author's punning title in both its meanings. *Ḥalbet al-kumyat* means 'the milking (i.e. pouring) of red wine' and 'the race track of the bay horse'. Strachan understood the reference to drinking but not to horse racing. The different meanings arise through the changes possible in vocalisation of the Arabic letters where consonants are fixed but vowels are altered by the presence or absence of additional markings. Strachan's failure to notice such a point hardly detracts from the substantial mastery of Arabic that he had gained by 1618 both in everyday speech and the reading of classical literature. In Dellavida's opinion Strachan:

> was able to grasp the meaning of many literary works which would have been far above the understanding of any of the few Europeans who were conversant with Arabic, even of the great Erpenius; and probably his knowledge of Persian was of the same standard. (Dellavida 1956: 72)

Strachan's Contribution to the European Study of Islam

During his sojourn in Baghdad after he had lost contact with his Islamic tutors, Strachan continued to buy religious texts. By that time it is clear that he had a considerable knowledge of the tenets of Islam. It is through his collection and interpretation of these texts that Strachan made his greatest contribution to the West's understanding of Muslim beliefs and culture. At the beginning of the seventeenth century, Christian Europe's study of Islam was predicated on the belief that it was a 'false religion'. The scholar's duty was to show that Christian 'truths' were superior to Muslim 'falsehoods'. In his debates with his tutors in the desert, Strachan adopted this stance on an intellectual level and had been faced with arguments of a sophisticated nature in return. To counter them he had to engage in serious study of Islam in its various forms. In order to acquire knowledge of the Sunni–Shi'a schism and the importance of the differences in the beliefs of their adherents, he studied their religious texts and later purchased a substantial number of books on these subjects. His surviving library contains nine on Islamic theology – three Sunni texts (Appendix: Strachan's Catalogue Mss Nos 23, 33, 55) and six Shi'a (Appendix: Strachan's Catalogue Mss Nos 11, 34, 38, 41, 43, 44). In addition, there are a further six on religious law and tradition – four Sunni (Appendix: Strachan's Catalogue Mss Nos 7, 18, 37, 52) and two Shi'a (Appendix: Strachan's Catalogue Mss Nos 34, 36). The book he bought in 1618 from Ibrāhīm ibn al-Ḥasan, which was the excuse he used to leave his wife and go to Baghdad, was the Shi'a prayer book, *aṣ-Ṣaḥīfa al-kāmila* (Appendix: Strachan's Catalogue Ms. No. 43). He must have been pleased with the scribe's work since he immediately gave him another major commission which Al-Ḥasan completed in June of that year. This second volume is a work of the Shi'a faith and is one of the most important that Strachan sent to Europe. *I'lām al-wará bi-a'lām al-hudá* ('Teaching human beings the signs of the right path') is a biography of Mohammed and the twelve Shi'a imams which explains the development of the doctrine of the temporary disappearance of the twelfth imam. It was written in the twelfth century by a Shi'a scholar, Raḍiyyaddīn al-Faḍl ibn al-Ḥasan at-Ṭabarsī, and manuscripts of the original text are extremely rare. Strachan's copy is one of four extant: the others are widely spread – in the British Library, India and Iran with none left in the Arab world (Appendix: Strachan's Catalogue Ms. No. 41). This book and Strachan's commentary on it were to have a major influence on Western scholars' understanding of Islam.

Along with the rest of his library, these religious texts languished in Rome waiting his return from the East. The Carmelite friars no doubt referred to them in the course of their language studies, but it was over half a century before they received any serious attention from scholars. Ludovico Marracci was a member of the Order of the Clerks Regular of the Mother of God of Lucca who learned Arabic in Rome, which by the middle of the seventeenth century had become a major European centre for learning Eastern languages. He became expert in the language and in 1671 published a much improved translation of the Bible into Arabic. Before that he had begun what turned out to be a life's study of Islam with the intention of refuting its teaching. He studied the Arabic texts which had been accumulated in Rome to advance his studies of Mohammed and the Qur'an. In his writings he mentions that he worked at the library of the College of St Pancratius of the Discalced Carmelites where Fr Vincent had deposited Strachan's library. In particular he used Strachan's copy of *I'lām al-hudá* in his most important work. In 1691, using the textual glosses that Strachan had provided, Marracci published an extensive and learned exposition and refutation of the theological system of Islam. In it he wrote a life of Mohammed, quoting extensively from *I'lām al-hudá* (Dellavida 1956: 90). Marracci followed this work in 1698 with much improved editions of the Qur'an in Arabic and Latin both of which relied on Strachan's work for insights into difficult passages.

These new publications marked decisive improvements in the knowledge of Islam in the West. Although Marracci's intention was to attack Islam from an intellectual point of view he did not, as earlier translators had done, distort meanings and his translation was soon recognised as the most accurate available. It was used by Protestant scholars as the basis for translations into French and German, although with a loss of accuracy from Marracci's Latin version (Glei and Tottoli 2016). George Sale did not acknowledge his source, but his English translation of the Qur'an (*The Koran, Commonly Called the Alcoran of Mohammed,* 1734) was based almost entirely on Marracci's work and became a standard reference throughout the English-speaking world. Edward Gibbon drew on Sale's publication for his commentaries on Islam and the Qur'an and other prominent figures including US President Thomas Jefferson are known to have possessed copies (Hayes 2004: 247).

Strachan never returned to Europe but through the rich legacy of his library and particularly the translations and commentaries which the books contain, he was able to have a profound influence on Western scholarship of Islam and Arabic language and literature, albeit delayed for more than half a century after his death.

CHAPTER NINE

The English East India Company

Saving William Nellson

Strachan left the Bedouin camp for the last time in 1618 fleeing from Emir Feyyād, impending circumcision and his new wife. It had been five years since he had taken his impulsive decision in Aix-en-Provence to travel east and in the intervening period he had visited many of the Orient's great cities, had learned the Turkish and Persian languages, and mastered Arabic, both demotic and classical. He had studied under eminent scholars who introduced him to a range of important works and helped him understand the subtleties of their texts. In the process he had absorbed a great deal on the culture and lifestyle of the Arab and Islamic worlds. In material terms, he had accumulated significant savings as well as having gathered an impressive collection of books which were to be of great intellectual value in the West. To a large extent he had accomplished what he had set out to do, and in his letter of 26 February 1615 to Dupuy he had written that it was his intention to return to Europe in two years' time. From Baghdad, to return by the way he had come, involved crossing the Great Syrian Desert, the territory controlled by Emir Feyyād. While he remained in Baghdad, Strachan was fully aware that the emir and the wife he had abandoned had not despaired of his return and were looking for an opportunity to reunite him with his Bedouin family. Merchants arriving from Aleppo and Damascus would have told him of the emir's continued vigilance. By June when the scribe al-Ḥasan had completed his second commission, Strachan had been in the city for about four months and had spent a significant sum on books as well as living expenses. It would have been clear to him that the longer it took Feyyād to lose interest the more depleted his savings would become. It made little sense for him to remain in Baghdad.

Given that the way westward was barred to him, he decided to travel east. He let it be known that he intended to visit the court of the Great Moghul emperor in Agra. From India it would be possible to return to Europe travelling on one of the merchant ships that the Portuguese, French, Dutch and English were sending on regular voyages to buy spices in the west coast ports of India. The most convenient route for him to take from Baghdad was by river and sea, down the Euphrates to Basra on the Persian Gulf and by sea-going dhow through the Gulf and across the open ocean. The timing of deep-sea sailings was governed by the seasons, and mariners relied on the Monsoon winds to help in crossing the Arabian Sea. The south-west

Monsoons come in the summer months of June and July, and by the time Strachan had received his second book from al-Ḥasan it was too late to start out in that year's sailing season. In order to follow his plan, he would have to wait until the following year. While he waited, another course of action presented itself.

On 13 October 1618 Henry Saville, an agent of the Levant and East India Companies based in Aleppo, wrote to his superiors in London recommending 'Strahanna, a Scotchman residing in Bagdad, a physician' as very fit to carry letters to Persia, an undertaking that he described as dangerous (Sainsbury 1870: 233). This is the first report of the relationship Strachan developed with the 'Turkey merchants', as the Englishmen trading in the East under the umbrella of the Levant Company were known. While in Constantinople it is possible that he met their ambassador, Paul Pindar, who was the last company appointment to the post. In 1620 King James was to select a court diplomat, Sir John Eyre, to replace him. From his coronation as king of England in 1603, James' foreign policy took a different direction to that of his predecessor. Whereas Elizabeth saw Spain as England's arch-enemy, James was trying to improve their relationship. To this end he began to distance himself from Spain's great enemy, the Ottoman Empire. His appointment of Eyre as ambassador was to ensure that England's attitude to Constantinople was not determined solely by the interests of the Levant Company. At the time of Strachan's visit to Constantinople in 1614, England had only mercantile relations with the Ottomans. Any earlier relationship that Strachan had developed with Pindar while in the capital would have helped him when he met with the Turkey merchants in Aleppo. As personal physician to Emir Feyyād, he was well known to all of the European merchants in the city but, after he fled to Baghdad, an incident occurred which forged a practical relationship with the English.

While in Baghdad, Strachan received an appeal from a Turkey merchant who had been imprisoned by the city governor. William Nellson had been apprehended, having in his possession letters that he was attempting to deliver to the agent of the English East India Company in Isfahan in Persia. It is possible that these commercial letters were in cipher, which would have added to Nellson's problems. He was travelling between the Ottoman and Persian Empires with no trade goods and was believed by the aga to be a spy. The war, which had been waging between the Ottomans and Iran for sixteen years, was reaching a dramatic conclusion and was going very badly for the Ottomans. Tension was high in Baghdad for good reason. The city was on the frontline of the conflict, and six years later the Safavid army captured it. The governor was not inclined to take risks regarding William Nellson, and the Englishman was on the point of being executed when Strachan intervened. The Scot was able to present the governor with a case for Nellson's innocence. The governor gave him a hearing, perhaps due to his status as a member of Emir Feyyād's divan, but it is likely that Strachan gave the aga a suitable 'present' to secure the merchant's release. On regaining his freedom, Nellson returned immediately to Aleppo.

As Strachan later explained in a letter to the governors of the company, dated March 1620, he had saved Nellson from being burned alive along with his letters (Yule 1888: 324–5). This was no idle claim as Ottoman practice in executions included flaying alive, boiling in oil and impalement. A public execution where a Frankish *kafir* spy was burned alive would not have seemed extreme to the citizens of Baghdad at any time, not least when they were concerned about the outcome of the war. On his return to Aleppo, it is easy to imagine Nellson's account of his arrest and rescue having a chastening effect on his colleagues. He also appears to have passed on a proposal from Strachan that the company could use the Scotsman to carry correspondence in such dangerous circumstances. As well as making this suggestion, Henry Saville's report to London can be viewed as an alert to the board of governors that they could not expect their Levant agents to make personal contact with East India Company merchants in Persia.

English Trading Companies in the East

The two organisations – the Levant Company and the East India Company – had been established under separate royal charters but had many investors in common. Sir Thomas Smythe, governor of the East India Company, had been governor of the Levant Company and was still a principal shareholder. Where possible it was expected that their merchants would cooperate but their interests did not coincide in every respect. In the late sixteenth century, as mercantile links between England and North Africa grew, merchants banded together and sought to regulate the new Mediterranean markets that were opening up. The Turkey Company was the first to form and received its charter from Queen Elizabeth in 1581. The Venice Company followed almost immediately in 1583 and the short-lived Barbary Company, which hoped to oversee trade in North Africa, in 1585. These royal charters expired in 1592, and all of these companies were superseded by the more enduringly successful Levant Company, whose charter gave it a monopoly of trade between England and the Mediterranean (Vitkus 2008: 20).

The company tried and generally failed to ensure that all English trading in the Mediterranean took place with its approval. There were simply too many adventurers in the region willing to defy the company's legal monopoly in pursuit of a profit. The founding members had been drawn largely from the trade guilds of the city of London and dealt with the commodities that their city guilds handled as monopolies. Individual guild members were able to buy trading rights from the company for a payment of £25 and some also gained official appointments as resident agents (MacLean 2011: 81–5). The wide variety of trade goods available – silks, cottons and spices from the East, currants from the Greek islands, opium from Turkey and a range of other exotic items – made it impossible for the guilds to control the growing trade. Other merchants came who were only tenuously associated with the chartered companies, but who involved themselves on the pretence that they were duly accredited. Added to this mix were those who, when the occasion presented itself, indulged in piracy under the guise of unauthorised privateering (Moshenska 2016: 241–2). This behaviour was entirely predictable given that the merchants' ships engaged in the Levant trade were heavily armed, typically with twenty to thirty heavy cannon, and obtained some members of their crews from the Fleet prison in London.

The Levant Company was already well established when in 1599 the East India Company received its royal charter to trade east of the Cape of Good Hope. The company structure was different to that of the Levant, in that it was created as a joint stock company. All of its traders were salaried employees and all trading profits were remitted to the shareholders in London (Robins 2012: 45). Using this organisational structure, the company directors hoped that they would maintain control and avoid the abuses of trade which prevailed in the Mediterranean. This was a forlorn hope. Representatives of the East India Company had personal backgrounds that were as varied as the Levant Company merchants, and company records show many examples of employees carrying out trading ventures for personal profit. In theory the two companies were not in competition since their charters covered geographically different regions, but as the East India Company developed it deprived the Levant Company in Aleppo of the profitable trade with the caravans from the east – a process that had been started by the Portuguese and Dutch. Using the sea route to England, the East India Company was able to cut the prices of pepper, mace, cloves, indigo and raw silk by almost two-thirds compared to the overland route via Aleppo (Mukherjee 1974: 393). Henry Saville's reluctance to have close dealings with men from the East India Company is understandable, even without considering the risk that Turkey merchants ran in attempting to cross from Ottoman to Safavid territory.

Despite the misgivings of the Turkey merchants, there was a distinct benefit to the company governors in London in dealing with the East India Company merchants through Aleppo. Thanks to the efficiency of the Venetian postal system, Henry Saville's letter would have reached Venice in a little over six weeks (Dursteler 2009: 601–23). The time taken for onward transmission to London was less certain but the governors would have been in receipt of the letter within weeks. In comparison, the ships of the East India Company took more than two years to return to their home port, even when their voyages were completed successfully (Robins 2012: 30). It took the company many years to improve on this. Although the frequency of sailings had greatly increased by the end of the eighteenth century, the journey time from India to London was still four months (Sutton 2000: 94).

This situation produced long delays in communication, ensuring that the governors had little control over the daily activities of their merchants.

Strachan was quick to understand the benefit to the East India Company of sending correspondence by way of the Levant agent in Aleppo, and in his letter of March 1620 he offered the governors the use of his contacts with Arab and Venetian merchants to ensure safe passage for mail and company employees between the Ottoman Empire and Persia (Yule 1888: 324–5). In the same letter Strachan also gave an account of his own dealings with the East India Company merchants in Isfahan. He had waited in Baghdad for a response from Henry Saville on his suggestion regarding transmission of letters. Although he continued buying books, there are signs that his money was beginning to run out. In the flyleaf of *Mu'allaqāt*, a book of ancient poetry, he wrote '*hunc librum propter raritatem et chari tatem venalem non repperit*' – 'On account of the rarity and value of this book he could not afford to buy it but had it copied' (Appendix: Strachan's Catalogue Ms. No. 53).

East India Company Trading Factory in Isfahan

By late spring of 1619, although he had still not received a reply from Saville, Strachan set out for Isfahan and presented himself to the company merchants. In his version of events, which he gave in his letter to the governor of the East India Company, he was travelling through Persia on his way to the court of the Great Moghul 'with good recommendations and faire expectations' and as a courtesy called on the company agent, Thomas Barker, and his fellow factors. It was they who persuaded him to stay and work for the company as a physician. Their offer was to provide him with board and lodging together with a modest payment which would be improved upon once they received authorisation from the governors in London (Yule 1888: 324–5). In addition, when he wished to move on to India, they would accommodate him on one of the company's ships. Clearly Strachan had called upon them seeking paid employment, concerned about the rate at which he was using up his savings. However, his situation was not as straightforward as he represented. An account survives in the East India Company archives of a meeting held on 20 June 1619 in Isfahan between the agent (president) Thomas Barker and two merchants, Edward Monnox and William Robins:

> [T]he President propounded, Whereas Mr. George Strachan a Scottish gent, is lately arrived from Bagdatt into this Cittie, and purposeth from hence to goe into India, whether it were not fitting and Civilitie in us for the time of his aboade here to proffer him a chamber and his diet in the Companys house and his passage . . . Hence to India uppon their next ship that shall heere arrive. (Yule 1888: 322)

The East India Company merchants arrived at this decision only after much discussion since they were not initially inclined to help Strachan. Their first instincts were not to trust him because of his background:

> This proposition being well debated and Severall objections made there unto, at first his religion . . . next his much breeding and long continuance in France whereby hee is become as well a French man as a Scotish man and verrie little or nothing at all an English man, nor as some suspect barely a good Subject unto the king of England, and therefore by entertaininge him into the Companyes House, he may get such insight into their busines that hereafter he may [prove?] verie prejudicial to the designes of our Honourable Imployers. (Yule 1888: 322)

Their first stated concern was that he was a Catholic, but this group of merchants was not wholly Protestant. One of their number, Robert Gifford, was also Catholic and, since he frequented the Carmelite convent in Isfahan, would have been known as such to the company (Yule 1888: 316) (Chapter 10). However, their concerns regarding Strachan's trustworthiness went beyond his confessional allegiance and nationality. Their account continues to explain the true reasoning behind their decision to offer him employment:

> Notwithstanding Such Severall objections it was for the reasons following generally concluded and resolved not onlie to receive him as a guest into the House, but to entertaine him as assistant to the Company. First that the Spanish Amb': [ambassador] hath been importune with him not only to accept of his House but also of some employment, which Hee intended to putt him in for the Service of the king of Spaine, which Service wee have iust Cause to Suspect is Cheifely by interceptence of our letters, by his meanes to have them translated, and to Come to the knowledge of the Contents of them, wherein Hee is so ingenious that our wrighting in Caracters would hardlie be concealed, where nowe our plaine Wrighting, neyther by the Ambassador nor yett by anie other in this Cuntrie (this gent onlie excepted), canne bee translated or anie way understood. (Yule 1888: 322)

The Spanish ambassador, Don Garcías de Silva y Figueroa, had arrived in Persia in October 1617 to conduct negotiations with Shah Abbas on behalf of King Philip the third of Spain and second of Portugal regarding an alliance against the Ottoman Empire. The shah had initiated the negotiations in 1601 by sending envoys to European courts with proposals for coordinated attacks on the Ottomans. In general the responses to Abbas' overtures were warm words but without any commitment to action. De Silva had left Spain on his mission to Persia in 1612, but had been detained in Goa until 1617 waiting for the shah to grant an audience. Even though the Spanish ambassador came in friendship, the shah saw him as a potential enemy and made him wait four years before he would allow him to enter Iran (Bull 1989: 164). Abbas viewed the Portuguese occupation of the forts on Hormuz and Kishm Islands in the Gulf as an insult to Persia.

The principal target of his displeasure was the Portuguese viceroy in Goa who governed his country's possessions in India, the Persian Gulf and the East Indies.

When de Silva was finally given permission to attend Abbas' court, the purpose of his mission was largely irrelevant. The Safavid army was on the eve of inflicting a crushing defeat on the Ottomans without any external help. The ambassador's account of his travels makes no mention of his having met George Strachan (de Silva y Figueroa 1667). However, he does mention having met Pietro Della Valle, the friend of the Scot. The Italian nobleman wrote about his discussions with Abbas regarding the trustworthiness of de Silva (Della Valle 1664: vol. 2, 16–17). The East India Company merchants' account of the Spanish ambassador's meeting with the French/Scotsman and the offer of employment is almost certainly true. Strachan's decision to seek employment with the English rather than de Silva can be explained by the fact that the ambassador was coming to the end of his stay in Persia. He left for Spain shortly afterwards.

The Englishmen's concern about Spanish knowledge of their affairs was due to more than secrecy regarding commercial matters. All trade in Persia was highly politicised, and the English later became involved in Persian affairs of state. As well as being a successful war leader, Shah Abbas was an astute politician and took a keen personal interest in the benefits which could be derived from the West. It was only by maintaining a tight personal control of all aspects of the affairs of his empire that Abbas had succeeded in rescuing it from ruin.

Shah Abbas I, 'The Great'

When George Strachan arrived in Isfahan, Shah Abbas had been on the throne for thirty years. This stability of government belies the fact that the start of his reign had been extremely turbulent. His father, Mohammad I, had been a weak ruler who was dominated by his queen, Khayr al-Nisa Begum, until she was assassinated by Qizilbash nobles. Mohammad was totally under their control for the rest of his ten-year rule. They formed the bulk of his army and he was unable to deal with their inter-factional fighting. His weakness was exploited by the Ottomans in the west and the Uzbeks in the east, both of whom had made significant inroads into Iranian territory. In 1587, when the Uzbeks launched another major incursion into Afghanistan, a leading Qizilbash nobleman, Murshid Qoli Khan, deposed Mohammed and installed Abbas as what he mistakenly believed would be his puppet shah. By playing the various factions of nobility against each other and, with the use of execution, assassination and exile, Abbas was able to remove the Qizilbash leaders including Murshid and his mother's murderers.

He confiscated their lands and replaced the nobles with governors whom he appointed from his corps of slaves (*ghulams*). All of these slaves were Circassians, Georgians or Armenians taken from the Caucasus, and they owed complete obedience to the shah. Thirty years earlier Abbas' grandfather, Tahmasp, had begun using Caucasian slaves in the imperial administration but only in relatively small numbers. Abbas increased their involvement greatly. Over the course of his forty-year reign, it has been estimated that he forcibly removed half a million Christians from the Caucasus to settle in the Persian heartland (Blow 2009: 99–103). Although the meaning of *ghulam* is 'slave', a more accurate interpretation would be 'servant of the shah'. As well as designating that their only loyalty was to the shah, this categorisation placed them outside the two other traditional groupings of Iranian society – those of the tribes of the Qizilbash and the feudal serfdom dominated by the old Persian nobility. By importing Caucasians in such large numbers, Abbas created a new stratum in Iranian society which came to be called 'the third force'. Most were required to convert to Shi'a Islam, and were settled throughout the country strengthening the Iranian economy as craftsmen, farmers, soldiers and government administrators. During his short reign Abbas' uncle, Ismail II, had tried to suppress the Shi'a branch of Islam in favour of Sunni. Abbas reversed this policy and strengthened the empire's identity as a Shi'a state by replacing those of the old administrative elite who were Sunni by *ghulams* who had converted to Shia. This action forced families, such as the Ḥusaynābādī, to flee Persia with the wealth of knowledge and literary works from which George Strachan was to benefit.

It was a deliberate policy of Abbas that senior positions in government and the army were occupied by his *ghulams*. By 1595 his commander-in-chief was a Georgian, Allahverdi Khan, who outranked the Qizilbash and Persian nobles. The new standing army of 40,000 *ghulams* contrasted with the feudal levies of the Turkmen tribesmen and the Iranian nobility. It was well trained, with a cavalry corps possibly numbering as many as 15,000, which made it the largest group of cavalry existing anywhere at that time. Abbas used it to enforce Qizilbash and Persian loyalty, causing the mounted Turkmen tribesmen of his father's and grandfather's armies to be relegated to auxiliary units of the main force of *ghulams*. From its beginning it was a gunpowder army equipped with and trained in using the most modern European firearms. The heavy and inaccurate harquebus, still used by the Ottoman armies, was replaced by the newly developed flintlock musket which was light enough to be used by the Persian cavalry. Abbas also formed a corps of musketeers 12,000 strong and complemented it with a separate artillery division of 500 cannon. The artillery of the time was still cumbersome, but that of Abbas' army had progressed beyond simple siege weapons and could be used on the battlefield. It took the shah almost ten years and a vast amount of wealth to create this new force. He was able to afford the expense, at least initially, through his appropriation of the provincial revenues which had previously been retained by the Qizilbash. His *ghulam* governors were appointed on merit and ran the empire in the interests of the shah more efficiently than the old aristocracy had done.

Once he had gained complete control of his father's diminished empire and had expanded and modernised his army, Abbas put the latter to the test. It was better equipped and trained than its enemies and was their superior in battle command structure. Abbas had been born in Afghanistan and it was a matter of personal pride that his home city of Herat be liberated. Starting in 1598 his first campaign was against the Uzbeks in the north-east. It took five years of continuous warfare before he had retaken most of the province. When his north-eastern frontier had been stabilised, he was able to turn his forces against the Ottomans. In 1590, as a young man of nineteen and while still struggling to establish control over the Qizilbash, he had been forced to sign the Treaty of Constantinople. In it he recognised Ottoman suzerainty over much of north-western Iran and the Caucasus. This had been necessary to ensure that he did not have to fight on two fronts while he was dealing with his Uzbek enemies. He had not forgotten this humiliation and in 1603, taking advantage of Ottoman weakness caused by military engagements in Europe and rebellions in Anatolia, he embarked on a ten-year campaign of re-conquest. He used his artillery to great effect against the Ottomans' fortified cities and retook his great grandfather, Ismail's, capital of Tabriz and most of the Caucasus. In 1612, by the Treaty of Nasuh Pasha, all territories ceded to the Ottomans by the Treaty of Constantinople were returned to Iran on condition of payment of an annual tribute. Abbas never paid the tribute, and the war resumed briefly in 1618 when the Ottoman army invaded Persia and was annihilated in an ambush at Sarab. It was at this point in the war that the unfortunate William Nellson fell into the hands of the governor of Baghdad and was accused of spying. Strachan's intervention to save the Turkey merchant and the Scot's subsequent journey to Isfahan came at a high point of Abbas' triumph against his greatest enemy. The shah's military ambitions, however, were not exhausted. Even before the battle at Sarab, Abbas had turned his attention to the Portuguese.

King Philip of Spain had sent his ambassador to Shah Abbas in 1612 during the period of peaceful relations between the Ottoman and Safavid Empires. Perhaps it was in part for this reason that, when de Silva arrived in the East, Abbas refused to grant him an audience. The viceroy of Goa, Don Jerónimo de Azevedo, entertained the ambassador for four years but the viceroy must have had misgivings regarding the mission to improve Spanish-Safavid relations. Azevedo was fully aware of the threat to the Portuguese from Abbas; the shah had retaken the island of Bahrain from them in 1602 as a precursor to launching his campaign against the Ottomans. Azevedo was energetic in his defence of Portuguese territory in the Gulf and feared that any agreement between Spain and Persia would be at the expense of Portugal, despite the union of their two crowns under King Philip. The king sent another viceroy, Don João Coutinho, to replace Azevedo and it may have been this act that persuaded Abbas, at last, to grant de Silva an audience. However, Safavid and Ottoman relations had begun to deteriorate by 1617 and the possibility of a Spanish attack in the west to divert the Ottomans may have been the stronger reason for the shah to meet the ambassador.

The Portuguese were right to wish to maintain their distance from Abbas since, shortly after the ambassador's arrival in Isfahan, Abbas' troops occupied the Persian coastal lands along the Arabian Sea declaring their ports open to all traders and thus breaking the monopoly that the Portuguese had established. This enabled Shah Abbas to control all Persian trade by sea as well as by land with the exception of one important route. The exception was the entrance to the Persian Gulf through the Straits of Hormuz. This continued to be controlled by the Portuguese from their forts on the islands of Hormuz and Kishm, which they had established early in the sixteenth century. The Spanish ambassador had come to negotiate on the proposal for an alliance against the Ottomans, but Abbas had little need of help from the West. Discussions were focused on the removal of Portuguese power from his domain. The shah did not need Spanish cooperation regarding the coastal ports: his army occupied them without difficulty, but he had no navy and the island forts were beyond his reach. De Silva had no remit from his king to discuss such concessions and left in 1619 with the matter unresolved.

It was less than a year before the ambassador's departure that George Strachan made his acquaintance. It is unlikely that the Scot became involved in any of the proceedings between de Silva and the Persian court, despite the obvious advantage to the Spaniard in having such a proficient linguist in his service. By the spring of 1619, when Strachan arrived in Isfahan, the Spanish mission had all but concluded with little success. Nevertheless, conversations with the ambassador would have provided him with insight into the workings of the divan. Strachan was later to form friendships with a number of the most senior members of the shah's court, and went on to use his influence with them and provincial governors to aid the East India Company as well as his friend Pietro Della Valle and himself (Chapters 10 and 11).

State Expenditure

Shah Abbas' interest in having complete control over trade was more than a matter of national pride. He needed as much revenue from taxation as was possible. The expense of the army and the cost of his wars were enormous: more than his income from the provinces could support. To these major projects, however, he had added another large expense. Before he launched his war against the Uzbeks in 1598, he decided to move his capital from Qazvin to Isfahan. The move was strategic, in that Isfahan was in the centre of the empire. Qazvin and the earlier capital of Tabriz were further west and had been vulnerable to Ottoman attack from the time of Tahmasp I. The Ottomans had occupied Tabriz for most of his father's reign and it had remained under their control until Abbas retook it in 1604. Isfahan was more secure from such attacks but was not impressive enough to be the empire's capital. Abbas set about planning a new city with magnificent public buildings – palaces, mosques, colleges, baths, souks, caravanserai and an enormous public square, Naghsh-i Jahan, which remains an outstanding example of Islamic art and architecture.

Isfahan was not simply his magnificent seat of government; it was also Abbas' intention that it should be a centre of Persia's most important commercial undertaking, the silk weaving industry. This was necessary to help provide the finances needed for the great public works that he was commissioning. In his early campaigns in the West, he had retaken almost the whole of the Caucasus but he was unsure that his army would be able to hold it against a renewed campaign by the Ottomans. In order to deprive Constantinople of the region's wealth, he decided to conduct a scorched earth policy which involved the mass removal of the population of the region to Persia as *ghulams*. Estimates of half a million people do not seem exaggerated. Whole communities were moved. Among them were many Armenians who had great skill in producing fine silk fabrics (Blow 2009: 99–103; Bournoutian 2002). Abbas moved the entire population of the Armenian town of Julfna to Isfahan and catered for them by creating a district known as New Julfna. There they were given a great degree of freedom. The Armenians replicated their original home including their silk works. As a means of ensuring their cooperation, exceptionally, Abbas allowed them to retain their Christian faith. He constructed churches for the community who kept their Orthodox priests and were presided over by their own resident metropolitan bishop. New Julfna grew as the centre of the silk industry which became a crown monopoly.

The highest quality of silk cloth was produced in Isfahan but other cities also had workshops for silk manufacture. Lars and Kirmān also excelled, particularly in carpets. In order to supply this greatly expanded industry with raw silk, Abbas regenerated much of the Caucasus and the lands along the Iranian shore of the Caspian Sea by encouraging peasant farmers to cultivate silk worms. In this way Iran was no longer dependent on imports of raw silk from India (Matthee 1999: 33–5). This was a reversal of his earlier policy of depopulation of the Caucasus, and is the clearest indication of his growing confidence in his military dominance of the newly reconquered lands. Abbas wanted the profits not only of the production of silk cloth but also of the revenues from its sale abroad. To achieve this he established his *ghulam* Armenians as the empire's silk merchants. They exported the high-quality silk fabrics, using connections with fellow Armenians in the Ottoman Empire, and took advantage of their established trade routes into Europe as far as the Armenian merchants' house in Amsterdam.

The financial return to Iran and the shah was in gold and silver, which he needed to support his pursuit of almost continuous wars and the creation of Isfahan as one of the great cities of the East. However, these were permanent drains on the imperial treasury that could not be met by the taxation of the provinces or the substantial revenue from the silk trade.

The Persian Empire had few natural deposits of precious metals: traditionally it had been reliant on the Ottomans and Russians as sources of bullion. The arms and munitions needed for the imperial army were bought with gold and silver, which Iran could only obtain by trade or occasionally as a result of pillage in war. Even before Abbas' reign the Safavid dynasty had been permanently short of money. Tahmasp and Mohammad were unable to pay their armies on a number of occasions which led to desertions and contributed to their military defeats. As a consequence of the shortage of the precious metals, throughout the two centuries of rule by the Safavid dynasty, the Persian currency was heavily debased, usually by the shah himself but sometimes by provincial governors who were allowed to run their own mints. Early in his reign Abbas had made a determined effort to reform Persia's coinage. In building his new standing army, he had recognised the danger in not having sound money with which to pay for adequate equipping of his troops. Despite being a 'slave' army, the provision of the advanced military equipment including muskets and gunpowder made it expensive to equip and maintain.

While he was still benefiting from the confiscation of the Qizilbash nobility's lands and wealth, the shah introduced a new silver coin, the 'abbasi'. From its inception the abbasi was not pure silver but was less debased than most Ottoman or Indian coins. The 'shahi' coin of his grandfather continued to be used in trade but at an exchange rate of five shahis to the abbasi, which reflected the degree to which the earlier currency had been debased. As his wars progressed and the demands on his treasury grew, despite his initial good intentions, Abbas resorted to debasing his new currency. By stages the silver content of abbasis was reduced until it was as little as 5 per cent by weight. The notoriety of the Persian currency was such that in the Treaty of Nasuh Pasha of 1612, by which the Ottomans recognised the return to Persian control of the Caucasus and north-western Iran, the clause introduced as a face saver for the Ottoman sultan stipulated that the tribute that Abbas was required to give annually was not expressed in coinage but as 500 bales of silk cloth. Debasement of currency did not improve under Abbas' successors as is illustrated by Sa'ib of Tabriz, the court poet of Abbas II, who thirty years after Abbas the Great's death wrote a poem containing the lines:

> Sa'ib, get yourself some iron object looking like a coin
> Because it's merchandise today that's close to gold. (Matthee 2012: 92)

The Silk Business is a 'Ready Money Trade'

Shah Abbas I's need for sound specie was such that he reissued an ineffective decree of Ismail II that all trade within Iran had to be conducted in Iranian currency. Traders arriving at Persian customs posts overland from Baghdad, in addition to being assessed for duty on their trade goods, were required to surrender their Ottoman and Western (usually Venetian) gold and silver coins for debased Persian abbasis. Understandably, they were particularly unhappy about this exchange of good coinage for bad. In the short term the effect of this unfair arrangement was that, whenever they could, merchants relied on barter and carried as little sound specie as possible, going to great lengths to hide it from the customs officials. The officials in turn became adept at finding the money since they received a portion of any they confiscated as a reward. In reaction to this behavior, Ottoman sultans banned the export of their currency to Iran. Merchants increased their use of letters of credit. In the longer term, this benefited the Armenian silk traders of Iran who, through their contacts with their countrymen in the Ottoman Empire, were able to conduct much of their business in this way. The commercial advantage which this gave the Armenians helped them further their domination of the land trade westwards from Persia. The letters of credit were redeemed with gold or silver – often Venetian ducats which were of greater purity than any of the Eastern empires' coinage. European money was so much more generally trusted that some Ottoman mints even produced copies of ducats as they were more readily accepted than their own coinage (Matthee 2012: 75–109).

Trade with Iran by sea equally was fully monetised. Russians sailed across the Caspian Sea to Iranian ports and knew that 'in order to buy raw silk one has to dispose of pure silver' (Soimonov 1762: 353–4). The merchants who arrived at the ports on the Arabian Sea understood that good gold and silver coins were essential for successful trade. Portuguese and Spanish reals were made from precious metals mined in their American and African colonies. Each year the Spanish sent a treasure fleet (the Manila Galleons) from Acapulco on the west coast of Central America to the Philippines to finance their trading in the Far East. One-third of the silver mined in America was used in Spanish and Portuguese trade with Asia (Furber 2004: 231). When Shah Abbas broke the Portuguese monopoly in 1618 the French Compagnie des Indes Orientales and the Dutch traders of the VOC set up their bases in Persia, fully aware of the terms of trade, and came with silver.

The English were equally aware. English travellers as early as the middle of the sixteenth century had reported that the silk business was a 'ready money trade' (Matthee 1999: 43). Despite this knowledge, when the East India Company arrived in 1618 to set up its trading station in Persia, it was short of the required specie. Like Iran, England had few natural sources of gold or silver at home. The directors of the East India Company believed that a similar approach to that of the Levant Company would be successful also in the Far East. The Levant Company operated largely by barter using English woollen cloth. Often this entailed extended voyages to various ports around the Mediterranean Sea. The cloth was exchanged for commodities such as grain in Egypt, which was in demand in the Greek islands, where the grain could be exchanged for currants which were sold in London (Moshenska 2016: 267–99). Such arrangements were not so successful east of the Cape of Good Hope.

The first venture of the East India Company in 1601 involved four ships sent to Indonesia to trade for spices. They went with silver and were successful in dealing with local rulers who were happy to receive better prices for their spices than they were given by the Dutch and Portuguese, who had monopolised the trade through force. English naval power was essential in achieving success since the European rivals used their superior sea power to crush competitors. In 1612 an East India Company fleet was victorious over the Portuguese in a naval encounter off Surat. It forced entry into the port and began the first English trading venture in India. Following this success the English ambassador, Sir Thomas Roe, was sent to the Moghul court in 1618 to negotiate a formal trading arrangement. This coincided with Shah Abbas' actions in breaking the Portuguese monopoly in access to his seaports. The East India Company trading venture in Iran was developed in tandem with that of Surat since the English had to protect their dealings in both locations against Portuguese opposition. India was well endowed with all kinds of desirable merchandise – silks, calicos, muslins and spices, but it had little need for European trade goods and, like Iran, all business in India required hard cash. By comparison, European trade with Persia was essentially limited to silk – raw fibres, fine finished fabrics and carpets.

In the early years of its formation, it was in the commercial interest of the East India Company to concentrate its limited resources of bullion on trade with the Spice Islands and India. When Strachan first met the East India Company merchants in Isfahan in the spring of 1619, they had been in the country for barely a year and were struggling to establish a profitable trading station. They faced myriad difficulties. As well as being under-financed, their language skills were poor and they had little understanding of the culture of Abbas' Iranian empire. In addition it was becoming clear that they were vulnerable to disease. Even before Strachan arrived, a number of the company servants had died and one of the merchants, Mr Robbins, succumbed to disease soon after Strachan met him. When Strachan presented himself as a physician and skilled linguist, the East India Company traders were keen to make use of him in both capacities, despite their reservations regarding his religion, nationality and questionable loyalty to King James.

The offer of employment had to be temporary and conditional on receiving approval from London. The smooth running of the trading station was handicapped by poor communications with the East India Company board in London. The time taken for letters to receive an answer was so protracted that paralysis in obtaining authorisation for any action was inevitable. The agent, Thomas Barker, was obliged to take decisions and seek approval afterwards. Three months after his initial report concerning George Strachan and before he had received any reply from the governors he dispatched another letter to London explaining that he had increased Strachan's salary from ten dollars a month (approximately equivalent to £24 sterling per annum) to 'twelve Tomans per annum' (one toman equalled 50 abbasis). In his letter Barker explained that Strachan was unhappy with the original terms and, in the short time he had been with them, he had become indispensable as a physician and a linguist. The death of William Robbins had clearly affected Barker who remarked on his demise in his letter. Barker himself was to die from disease that winter. Having a doctor who spoke their language must have given them some reassurance, but also all the Englishmen in Isfahan had found Strachan's help as an interpreter to be invaluable (Yule 1888: 323–4). The relationship between the Scotsman and the English merchants should have been to the advantage of all, but the English were at odds with each other and, as their arguments became more heated and personal, they began to denounce one another in correspondence with the governors in London. In the febrile atmosphere of a failing trading station, it did not take long for Strachan to become the target of personal criticism.

CHAPTER TEN

'Stracan our Infernall Phesition'

Mistrust

The company of Englishmen that George Strachan had joined in Persia was a motley collection of men of good character and education, mixed with others of little education who included the scourings of the Fleet prison. They had been drawn together by the common ambition of making their fortunes in the East. As company servants, they all received a salary which varied according to their responsibilities, but in each case would never have made them rich. To become wealthy, they had to engage in trade for themselves; this was in addition to any work they did for the company. The original intention of the East India Company's incorporation as a joint stock company was that its servants should not trade on their own behalf. Given the circumstances, this condition was impossible to enforce. In the first half century of the East India Company's existence, each trading voyage to the East was an individual investment with its profits or losses being assigned to the investors who backed it. This meant that investors in successful voyages made great profits, but when ships did not return their investors lost everything. The position of company servants abroad was equally financially precarious, and it was only natural that they should look out for themselves by indulging in private trading. In 1657 the company changed its investment arrangements by producing annual accounts and allocating the risks and profits arising from its enterprises among all the investors. At the same time, the futility of trying to prevent its servants from engaging in private trade was recognised, and they were formally allowed to conduct trade within the country and retain any profit they made; however, trading in goods to be shipped to England was strictly forbidden (Robins 2012: 24–5).

In 1619 the company trading station in Persia was beset with the problems these early restrictions caused and it was the job of the company president, Mr Barker, to keep control of private trading. Unsurprisingly, this led to conflict. The letters preserved in the archives of the East India Company throw some light on the troubled relationships which had grown there. It appears that when Strachan first arrived in Isfahan, the English merchants were still trying to secure their first successful trade on behalf of the company. An earlier failed attempt was the cause of acrimony among the traders. An account of a remembrance (memorandum) dated 16 May 1619 survives, in which Edward Monnox ordered Robert Jeffris and other company servants to account for 28 'baftas' (lengths of cloth) costing 450 shahis which were intended, along with other items, as a present for a Persian official named Sherary. The account goes on to record that Jeffris claims to have submitted a copy of the note drawn up to show that the gift had been delivered on 15 January 1619 (Yule 1888: 324). It would appear that either the bribe did not have the desired effect of facilitating a trade or, as Edward Monnox believed, Robert Jeffris had pocketed the present himself. Turning a blind eye to company servants indulging in private trading was common practice, but theft of company property was not. Without firm proof, little could be done against Jeffris and the others involved, but an atmosphere of mistrust had been created between the senior members of the team and the other factors. Strachan appears to have viewed Barker and Monnox as gentlemen with whom he could associate. Most of the other members of the team were of lower social status, less well educated and, as was the case with Eduard Patten and John Hawkyns, 'two Runnawayes from the ffleet', known to be criminals (Yule 1888: 327). His social standing separated Strachan from all the less reputable members of the trading station. In his capacity as translator, Strachan was required to be present at all important meetings with Persian officials and merchants. The opportunity for Jeffris and others to purloin 'gifts' intended for such officials would have disappeared unless they managed to gain his cooperation. From later events, it is clear that he did not become one of their group.

Seizing the Initiative

Barker's letter of 16 October 1619, in which he asked Sir Thomas Smyth, East India Company governor in London, to approve his decision to increase Strachan's pay 'in regard of his profession which is of Such necessary Consequence in these unhealthful Climates and for Sondrye other reasons' (Yule 1888: 324), is the last one surviving that the agent in Isfahan wrote. He died there on 30 November that year. Due to the time taken for news to reach London, there was a delay in appointing a successor. During the hiatus, Strachan took it upon himself to write to Smyth on 25 March 1620 to ask for a pay rise (Yule 1888: 324–5). His letter is informative of his actions during the nine months he had been in Persia and throws light on his plans for the future:

> Right Worshipfull Sir, It is not unknowne to your worshipp as I esteeme nor unto the rest of the honble: Company in what State and intention I came the last yeare into this Countrye and upon what Conditions and hopes I was retayned by the deceased Tho: Barker and the rest of the part of the Honble: Companies factors heere much against my owne intentions or desires, for haveing seen all Turkye and the most part of Arabia these seaven years nowe past and having learned the languadge

> I was passing into India to the Court of the Great Mogoule, with good recommendations and fayre expectations, but after that the aforesaid gentleman had with many reasons declared unto me the honnor and utilitie which I might [have] of the Honble Company yf I would accept of their service rather to spend my age in followinge of forreigne princes, I was persuaded by their Courteous offers to tarrye with them heere till I could knowe the Honble: Companies will in that respect.

The tone of his letter is in keeping with the nature of the man. Although it is respectful, it is confident, even boastful. His claim of having seen 'all Turkye and the most part of Arabia' is exaggeration. He had visited Constantinople, coastal Asia Minor, Palestine and the Levant but not Anatolia. He had seen much of Syria and Iraq, but it seems unlikely that he managed to visit the Arabian Peninsula or Egypt. Nevertheless, he had visited more of the Middle East than most Europeans. His claim, that he was reluctant to delay his journey to India but was persuaded by Barker to stay for the benefit of the East India Company, again is stretching the truth. Given that the context of the letter was a request for a pay rise, such claims are to be expected. More importantly, he opened his letter by reminding the governor of the service he had already given the company.

Thomas Barker appears to have kept him in his confidence regarding his communications with London, and the next part of his letter contains a gentle rebuke to the governors for the lack of a decision regarding earlier requests:

> And therefore at this present have taken the bouldnes with the Same, most humbly intreatinge him as head and governour of all that honnorable bodye after dewe consideration of my quallitie and service which I am able and willing to doe and performe for the Honble: Companies service to let me have an answere of what I can have and hope for of them yearely that I may the more deliberattely and contentedly continue in theire Honours Service wherein [?] honest men are dayly made riche . . .

The implication is clear: he will leave their service unless he receives a positive response to his request for an improved salary. The governors in London were unused to such demands. Employees in the East were dependent on the company for repatriation and once in post they had no real leverage over pay. They had to accept their conditions or be stranded. This was not the case with Strachan, and by remarking that honest men were making their fortunes in the company's service he was indicating that a token increase would not be acceptable. He was looking for a speedy reply given that nine months had passed since his original conversation with Barker, when they had agreed a temporary payment.

In the next section of the letter he reminded Smyth of the benefits he had already brought to the company:

> [I]t is well knowne to all those which be heere that laying aside the physick which is the principall cause of my entertainment not only I can serve them much by my languadge in this place but alsoe by the friendship which I have with the Arabian and Venetian marchants in Babylon [Baghdad] and Aleppo, and may cause thence letters to be safely conveyed to the Consull at Aleppo with easie expenses and without danger as divers tymes heretofore I have done and nowe this present packet by my means is sent by the said way of Babylon.

Strachan had spent the nine months since leaving Baghdad acting as the East India Company's doctor in Persia and facilitating the transmission of letters via Aleppo to and from the company headquarters in London. He continued to justify his request for an increase by suggesting other ways in which he could be of service to the company:

> Yea I may alsoe finde favour to cause the English passe safely through these Countryes when occasion shall offer that any should take theire waye thence by land, as I did faithfully and freely now two years agoe, in the person of William Nellson, the which if he had not found me at Babylon had assuredly bine burnt with his letters. And finallye I can serve you as well as any other in chooseing and buying of all such drugs which this countrye can affoarde.

His offer to extend the use of his contacts among Arab and Venetian merchants to bring company servants overland rather than by the long sea route via the Cape had clear benefits for the company and did not involve any activity that Strachan had not already engaged with. The offer of sourcing and buying drugs for export is a new departure and would have involved him in trade. As a gentleman sensitive about his status, this would not have come easily but shows that Strachan was thinking of gaining financially as well as intellectually from his journey to the East. Since he left the service of Emir Feyyād, his finances had been depleted to the point where he had stopped buying books.

The suggestion that he source drugs for the company was an intelligent one under the circumstances. Practising as a physician, he was aware of the range of medicines available in the eastern markets and of their uses. Taking advantage of this expertise would not have placed him in competition with any of the East India Company merchants dealing in silk and furthermore would open up a largely unexplored market to European trade. Spanish explorers had seen the value of new medicines brought from the Americas to Europe. Cinchona and guaiacum were used in the treatment of illness, but others, such as tobacco and chocolate, although they were used for recreational purposes, had first been promoted as medicines. The Levant Company was also having success with the importation of coffee and although its trade in opium from Turkey was small, the drug was much later to play a pivotal role in the profitability of the East India Company's trade with China. Strachan was suggesting to the governors in London that the drugs available in the East could be an unexploited source of profit. With these carrots dangled in addition to the undoubted service he was already providing as physician, interpreter and postmaster, Strachan felt that his true worth to the company was considerably more than the few abbasis that he was receiving:

> And in consideration of all the aforesaid I have demanded and hope to obtayne of the honourable Companie 100 pounds per Annum for all entertainement and charges to be paid me yearely here, whereupon I beseech your worship to Cause me to have an answere of the Honble Company by the ffirst letters that thereby I may be resolved what to doe, for if it shall not please them to honnor me with that answere I shall then take their sylence for a direct distast and Soe continew my begun Voyadge whether it shall please god to direct me. Thus praying your worshipp to pardon me if too rashly I have enterprized to importune him with these few lynes I comitt him humbly unto the protection of the Almightie from whom I devoutly wish unto your worshipp all prosperitie and felicitie.
> Your worshipps servant at Command
> George Strachan
> Spahan the 25th March 1620

By summer of the same year, the governors had responded by giving Strachan the salary he had requested. Apparently they decided that they could not afford to lose his services. Much as he was needed as a physician, it is likely that their real concern was the loss of the facility of communicating more speedily with their agent in Persia. It had taken only a few months between the death of Robert Barker and the governors' notifying Edward Monnox of his appointment as replacement agent. If they had had to rely on communications by the company ships, it could have taken as long as two years following the death before the new agent could have been in post. This circumstance alone could have been sufficient to persuade them to agree to his demands, although Strachan's other arguments must have carried weight. There was clearly a need for a doctor to tend to their servants, and Strachan's skill with languages had already been vouched for by Barker. Persia was not like the eastern Mediterranean or Aleppo. There was no lingua franca: trade was conducted in the two principal languages of the country, Farsi and Arabic. The English merchants had been at a severe disadvantage when they were without a competent interpreter whom they could trust. There is no evidence, however, that the East India Company governors took up Strachan's suggestions regarding sending personnel via Aleppo or sourcing drugs for the English market.

Malaria

Before Strachan received a reply to his request for a raise in salary, there was a significant change in his health. Shortly after writing his letter, he travelled to Shiraz in the company of Edward Monnox, to assist the English factor, William Bell, in arranging trades. Shiraz is a large city in the south-west of Iran on the road from Isfahan to Baghdad. It was the seat of the governor of the southern provinces of Iran and had a thriving silk industry. Bell was the East India Company man based there, entrusted with buying suitable merchandise for dispatch to England. The new agent, Edward Monnox, had instructed both Bell and Strachan to carry out private trading on his behalf. On 16 May 1620 Monnox wrote to another member of the group:

> Together with the rest of the goods mentioned in a Remembraunce left with your consignee unto Mr. Bell and Mr. Strachan and your self. I deseir both you and them to procure Sale for them, to my best advantage and investment, and retorne thereof to be made according to my forementioned writting. And not only of that but of all other monyes, and goods of myne, which shall Acrrewe unto me of right. (Yule 1888: 326)

Strachan was unable to act on Monnox's behalf on this occasion as he was suffering from a severe attack of malaria. William Bell advised Monnox of this in a note dated 8 May 1620:

> This bearer, Mr Strachan, since your departure, hath been visited with a voyalent burning feavour, and hath had ffitts already, which hath much weakened him, and hee much feareth if he should stay heere it would cost him his life, for he hath been very grevously handled, what having the Company of Signor Alviso Parent is determyned to depart this night towards Spahan, where he hopes to recover his health, that being a more wholesome ayre than this, espetially att this tyme of yeare, which I willingly consented unto, for his health's sake, because he could not now assist me in my bussines, being soe sick. (Yule 1888: 325–6)

Bell's description of the onset of fits (seizures) is indicative of a severe attack of malaria. Strachan was in his late forties and would have been seriously weakened by the disease. Medical practitioners of the time believed that the illness was caused by breathing bad air (*mal aria* in early Italian) and Strachan was anxious to remove himself from what he believed to be the 'bad air' of Shiraz as quickly as possible. The nature of the malarial parasite and the key part played by mosquitoes in its spread was unknown until the late nineteenth century. Recent research has thrown further light on the behaviour of the parasite. Once a victim is infected, the malaria parasite remains inside the host but changes from sexual to asexual reproduction.

In doing so, the initial vigorous attack is replaced by a quiescence which removes the distressing symptoms and gives the impression that the disease has gone. Whenever the parasite senses that its host is weakened and could die, it reverts to sexual reproduction which produces the symptoms of malarial fever. The parasite appears to do this to ensure its own survival, since it can only be transmitted to a new host in its sexually reproductive form (Edinburgh and Toronto Universities Study 2018).

Leaving Shiraz would have made little difference to Strachan's health. Breathing 'a more wholesome ayre' would have made no difference. Shiraz and Isfahan have similar climates – both are situated almost 5,000 ft above sea level and each has the reputation of being a 'healthy' part of the country. Shah Abbas' father, Mohamed, was raised in Shiraz in the belief that its temperate climate would improve his poor health. Strachan's insistence on leaving Shiraz is likely to have been influenced by the paucity of medical care that he was receiving there. Signor Alviso Parent, who accompanied him to Isfahan, as his name suggests, was a Portuguese Catholic who would have seen it as his Christian duty to take his co-religionist to the convent of the Discalced Carmelites in Isfahan. There he would receive the care he needed to convalesce. It was part of the friars' duties to nurse the sick. The standard treatment for malaria at the time, which proved to be relatively effective, was rest and, where possible, the administration of drinks made from an infusion of extract of cinchona bark (quinine) which Strachan would have possessed in his medicine chest. On this occasion he did recover and it is likely he believed that he was cured, but he still harboured the parasite and on a number of occasions the fever recurred.

Robert Jeffris' Bile

Over the course of 1620, the acrimony between Robert Jeffris and senior members of the agency, which had originated the previous year, intensified. Many of the surviving records on the subject derive from Jeffris himself, but even so, the impression given is that he was a particularly unpleasant person. There is evidence that he was a sneak thief, liar, bully and, without doubt, perfidious in his dealings with his colleagues. Following the appointment of Edward Monnox as replacement for the deceased Barker, Jeffris wrote to London to blacken Monnox's character. His attacks were contained in a long letter in which he made twelve accusations of wrongdoing against him. He also attempted to incriminate the chaplain, Mr Cardowe, and George Strachan. In his first accusation, he claimed that Monnox and Strachan had cheated the company and hidden the fraud using false accounting. Strachan had attended Thomas Barker as a physician during his last illness, and Monnox paid him for his ministrations but recorded the small sum involved as part of a donation to the poor made on the death of Barker. Jeffris wrote that he had to inform the governors of this because he could not take the matter up with the agent, since on previous occasions when he had tried to report the wrongdoing of others to him he had been rebuffed and humiliated:

> When I tell the Agent privatly of such and such burthensome servants unto the ffactory, he declares yt publiquely either at dinner or supper that I seeke to cleanse the Company from their service, thinking thereby to procure their hatred of me: but for all I thanke God I am armed with patience. (Yule 1888: 327)

Jeffris did not need Monnox to turn the others against him, as he managed that by himself without any help. In a note of a meeting on 14 August, it was recorded that Jeffris had entered the company house to find the doorkeeper and two other junior company servants drunk. Edward Patten was attempting to support his colleague 'whom Wyne bereaved of ffootmanship' when Jeffris set on him with his riding crop. Patten was angry but did not retaliate other than to tell Jeffris that if he did it again he would strike back since Jeffris had no authority over him (Yule 1888: 327). It would appear that Jeffris was disliked and distrusted by all the East India Company employees in Isfahan, both senior and junior.

Being friendless, he tried to ingratiate himself with the governors in London at the expense of his colleagues in Persia. In his long letter to London, Jeffris' second accusation concerned private trading which he described as dealing in contraband goods:

> No. II is his publique private trade formally advised by my letters concerning 30 bales Baftaes and Shashes and 25 Bales of Gumlacke newly arrived heere, which is carried so privately from my notice as may bee. And our Scotsman is his ffactor for underhand dealing in this busines de contrabanda. (Yule 1888: 327)

The route taken by the company's ships sailing from London included calling at the port of Jasques (Bandar Jask) in Persia as well as Surat and the Spice Islands. They delivered trade goods and bullion at the same time as collecting cargoes for return to England. Persia imported fine Indian cottons and spices and, while they remained in the East, the ships engaged in this trade exchanging cargoes between Surat and Jasques in the same way as the Levant Company traded in the Mediterranean. Jeffris' accusation was that Monnox had used a company ship to import these materials from Surat on his own account, a practice that went against a company rule that was honoured more in the breach than the observance. His eighth accusation includes a claim that Strachan charged the company for the use of his books 'for bettering Mr. Cardowes studies'. It would appear that Strachan had undertaken to teach the clergyman Arabic or possibly Farsi using his books – most likely his copy of the New Testament which had been his own early primer. His student would have taken offence at using works on Qur'anic and Islamic studies but not, if Jeffris' report of the clergyman's character is to be believed, *Ḥalbet al-kumyaṭ*, the book on the pleasures of drinking and gambling that Strachan had bought in Baghdad. Jeffris' complaint was mainly directed against Reverend Cardowe who 'hath not studied after the rate of 1 per cent, (unless it be in Tobacco and wyne, and Sleepe) in the books' (Yule 1888: 327).

Jeffris' attack on Strachan was much more serious. He was envious of his greatly increased salary and became vindictive towards the Scotsman. Although he did not include it in his first letter to London, he started a rumour that proved extremely injurious to Strachan. He suggested that the company doctor had poisoned Robert Barker and a merchant named William Rynns, who had also died recently. As is the case with most rumours, there were some among the East India Company merchants who were prepared to give credence to the accusation, despite their dislike of its source. The death rate from disease among the merchants was high. There had been deaths before Strachan's arrival in Isfahan, and health problems continued to plague the company throughout more than two centuries of conducting business in the East: during that time, over half the company's European servants died in post from endemic diseases (Robins 2012: 30). Nevertheless, the English merchants in Persia, like other small closed communities, appear to have needed a scapegoat for events they could not otherwise explain, and Strachan was an outsider. When he first arrived from Baghdad the merchants had reacted to him with suspicion, which they justified by the fact that he was a Catholic and, because of his education, he was 'as much a French man as a Scotish man and verrie little or nothing at all an English man' (Yule 1888: 322). Despite the total absence of proof, the rumour of the poisonings took hold and, as one can imagine, Strachan was greatly concerned. It touched on his honour as a gentleman, and he was eventually driven to put a formal request in writing to Monnox:

> I beseech you as before in the King and Companys name to free me of this Ignominye and Shame, by correcting and restrayning this evile and envious man, or if his malicious minde and detracting toungue Cannot be bridled, governed or restrayned, to give me licence to goe out of this house and permit that I may live in peace and honour amongst strangers seeing I cannot find them amongst my Countrymen. (Yule 1888: 328)

It was in Edward Monnox's interests to get rid of Jeffris. On 21 August 1620, the week after the assault on Edward Patten, Monnox held a consultation with other company merchants after which both Jeffris and Strachan were dismissed (Yule 1888: 328). The reasons for the decision are not stated. The note in the East India Company records is only of the decision without any detail. Jeffris claimed in later correspondence with London that Strachan had fabricated evidence against him. He claimed that the physician had presented Monnox with a counterfeit document which was intended to incriminate him:

> The forgatory ridiculous, unauthorised treason (as they would have it) was exhibited by Stracan (with other liberties of his own invention) unto E. Monnox on Sunday, the 17th of August last the some thereof was: That in ffebruary 1619 in Xiras [Shiraz] I should tell one Giles Gonsalves (a Portugall) that our vertuous Queen Ann (of happie memory) died a Catholicke, And that our hopefull prince Charles was tutored in the Papist religion. (Yule 1888: 332)

The document was a copy of a gazette (called *Gazita* by Jeffris) which appears to have been circulating among the Portuguese community. The Carmelite convent in Isfahan possessed a printing press and the gazette was probably produced there (Chapter 11). The clergyman, Mr Cardowe, testified that the defamation of the late queen and Prince Charles amounted to treason and that the gazette was proof of Jeffris' guilt. The issue was politically sensitive. Negotiations had been in train for some years regarding a marriage between the prince and the Infanta, Maria Anna, daughter of Philip III/II of Spain and Portugal. A key part of the negotiations involved the conversion of the prince to Catholicism before the match could take place. Such a move was anathema to most Protestants in England. Those close to the court would have known that the recently deceased Queen Anne had converted to Catholicism some years before. The claim that Prince Charles was tutored in the Papist religion was untrue. The infant Prince Charles had been raised by his Catholic mother until she left with the king for England in 1603. He was too sickly to travel and was left in Scotland under the care of the Lord Chancellor, Alexander Seton, who was Catholic. However, the following year, at the age of three, Charles was sent south and it is unlikely he was under any Catholic influence in England since his mother's Jesuit confessor, Robert Abercrombie, was exiled following the Gunpowder Plot in 1605. After the age of five, Charles' only close dealings with a Catholic were with his mother who was discreet at court about her faith. It is likely that Robert Jeffris and the other traders in Persia knew of rumours regarding the queen, but there was one man who knew the truth and that was George Strachan. He had been aware of the situation at the Scottish court in 1601 during his visit with Fr Abercrombie, the priest who had inducted the queen into the Church.

The production of this document was so fortuitous to Monnox's interests that it is almost impossible not to believe Jeffris' claim of innocence. But that could only be the case if Strachan had been involved in the forgery and this is difficult to reconcile with his dismissal by the same tribunal. If the gazette was a genuine document, it is understandable that Strachan should have been the person to discover it since he had associated with the Carmelite convent from his arrival in Isfahan. The document's link with Jeffris as the source of the published information could only be true if the trader had been indiscreet to a foolish degree. Given his behaviour, this is not impossible to believe. Nevertheless, his claim that the agent, the minister and the physician had conspired to libel him is credible. In a letter which he later wrote to the governors from Surat he put a forceful case that the real reason for his treatment at the hands of Monnox was:

> [M]y discovering the unreasonable, inconscionable corruption of Mr. Monnox in certayne percells of iniquitie, hath been (with the dispensation of the divell) a tryall treckerye begotten against me by our criticall agent Mr. Monnox, our carnall minister Mr. Cardo and Stracan our infernall phesition, the world, the flesh and divell. (Yule 1888: 332)

He weakened his case, however, by descending into almost hysterical language in his diatribe against his three oppressors. His description of Strachan is typical of his rants against Monnox and Cardowe:

> And Stracan our Antechristian Phesitian, for his filattering, lying, dissimulation, inconscionable stores of purloynment, with his tentarhookes of deere penniworthes of plaisters and purges, sowing dissention in the ffactory, his scandalous reports of poyzoning the Company's servants as the late agen and William Ryns, his dicouering of all the passages of our business to the ffryers in Espahan through his confession and disloyall service to the Company, intercepting of their letters, How can he be otherwise, being married to a More in Arabia, from whom he tooke his runnagate raunge, leaving wife and family to prosecute the divells commission in doing evil; continewally despizeth his owne country, and yts church, And confesseth to haue the dispensation of the Pope to dissemble his Religion in all his Pilgrymage.

He inflated his claim that Strachan was a poisoner to a preposterous level by 'discovering' in Surat two men, an Italian named Pietro Cheuart and another European, Estefano de Sant Jaques, who claimed that they had heard two Portuguese friars say that Strachan had told them that he had poisoned his English charges. They were persuaded by Jeffris to put this convoluted nonsense in writing, and he enclosed a copy of their affidavit in his letter to the company governors. There is no record of any direct response from London but damage had been done to the reputations of Monnox, Cardowe and Strachan.

Although the incident had descended into farce and Strachan had been dismissed from the company's employment, Monnox's treatment of the two men is indicative of where his sympathies lay. Immediately following his decision to dismiss, the agent took Jeffris into custody and escorted him to Jasques where he placed him on board the company ship, *The London*, informing the ship's company that Jeffris was 'A Prisoner for the Kinge' (Yule 1888: 330). The ship sailed for Surat where Jeffris was put ashore. Strachan was not arrested, despite the charge against him being one of murder, and he was able to take up residence with the Carmelites in Isfahan in the way he had requested in his earlier letter to Monnox. The company sent a replacement doctor from England, Thomas Quince

(McRoberts 1952: 125): the stigma appears to have remained on Strachan of being a poisoner or, if not, at least an incompetent doctor. The ramifications of the affair continued for the rest of 1620. From Surat Jeffris was able to influence the governors in London against Monnox who was removed from the post. In December Jeffris was appointed in his place but died on the voyage from Surat on his way to take up the position of agent. William Bell, the merchant in Shiraz, became agent for Persia but could not stand up to the more forceful Monnox, who remained to direct affairs despite having no official status (Forster 1906: vii).

Capture of Hormuz

Strachan stayed with the friars for some time before being re-employed by the East India Company, which happened through Monnox's influence. The date of his resumption of working with the East India Company is not recorded. In his biography Professor Dellavida states that Strachan stayed for a year before returning to work with the English company. Pietro Della Valle wrote that Strachan stayed with the Carmelites for a few months and this appears more likely, since if he had remained at the monastery for a full year he would not have taken part in what was to be the most momentous enterprise the East India Company was involved with in Iran. Less than three months after Strachan's dismissal, political events arose that required his recall. Edward Monnox needed his services as an interpreter.

Shah Abbas was planning to remove the Portuguese from the forts from which they controlled the Straits of Hormuz. In an attempt to regain their monopoly, Portuguese ships had continued to attack European vessels using Persian ports. Abbas lacked a fleet able to defeat them, and wanted the use of East India Company ships and crews to help in a campaign to take the forts. The English had proved their naval prowess in 1618, when they defeated a Portuguese flotilla that had attempted to prevent them entering Surat, and again in 1620 when Portuguese ships had attacked the English base at Jasques (Sykes 1915: 278). However, the Persians had difficulty in persuading the English to lend them their assistance in the dangerous enterprise. Negotiations were conducted between the representatives of the company and the *ghulam* general, Imam-Quli Khan, who was the son of the commander-in-chief of the army, Allahverdi Khan. The Georgian had shown himself to be as capable as his father, and the shah appointed him governor of the empire's southern provinces and entrusted him with the task of taking the Portuguese forts. The negotiations proved to be complex and prolonged during which both sides needed competent interpreters. There are reasons to believe that the East India Company re-employed Strachan to help in this. Quli Khan may not have had an equally skilled interpreter and perhaps used the Scotsman as well. If this was the case, it would explain why later he was held in high regard by senior Persian officials (Chapter 11).

There is a remote possibility, however, that Quli Khan had his own English interpreter. A number of Englishmen had arrived in Persia in 1598. They were led by three brothers, Thomas, Anthony and Robert Sherley, who claimed to have been sent as ambassadors by the Earl of Essex. A highly coloured account of their adventures was published in England in 1607, which perpetuated the lies and exaggerations that the brothers spread (Ross 1933: 91–7). In that account, the English brothers were responsible for reforming and training Shah Abbas' army in the style of English troops, which enabled the Safavids to defeat the Ottomans. This version of events is at odds with the known facts. The brothers were footloose adventurers who had no authority to speak for anyone. When they arrived in Persia, the shah had already spent ten years forming and training his *ghulam* army and was about to embark on his successful campaign in Afghanistan against the Uzbeks.

In his account, Thomas Sherley also claimed that he had convinced Abbas to send him as ambassador to European courts to organise a combined European and Persian attack on the Ottomans. Abbas did send an embassy, but his appointed ambassador was a Persian diplomat, Huseyn Ali Beg, and Thomas and Anthony Sherley accompanied him as guides. In an account of the embassy written by one of the ambassador's Persian secretaries, Uruch Beg, Anthony Sherley was described as a charlatan, liar and murderer (Beg 1604). Uruch Beg could also have added 'thief' to the description since, among other items, he stole the shah's gift to the pope. The two brothers never returned to Persia but their younger sibling, Robert, remained behind along with a number of other Englishmen from the original party. Robert joined the shah's army and, by his own account, helped Allahverdi Khan to continue with his reforms and went on to distinguish himself in battle. It is possible that there is some truth in this account since he was able to marry a Circassian noblewoman and return to Europe in 1615. None of the Sherley brothers were in Persia at the time of Quli Khan's discussions with the East India Company. It is possible, although unlikely, that a member of the original party who remained behind was still available to act as interpreter. Robert returned to Persia in 1628 with Sir Dodmore Cotton, the first English ambassador to the Persian divan; both were intent on encouraging Anglo-Persian trade but Sherley died shortly after arrival and, with the death of Shah Abbas the following year, East India Company trading interests never developed as strongly in Persia as they were to do in India (Chaudhuri 1965: 64).

Whoever acted as interpreter in the negotiations must have been skilled in both language and diplomacy. There were many months of negotiation and preparation before the English fleet sailed to Kuhestak. A year earlier, Quli Khan's first approach to the English merchants had been threatening. He demanded that the English ships with their crews support the Persian army in their fight with the Portuguese, otherwise their licence to trade in Persia would be revoked and their stores of silk in transit would be confiscated. The merchants were alarmed and wanted to cooperate, but the seamen refused. Faced with an impasse the Georgian general had to seek a solution through mutual benefit. The eventual agreement included a series of concessions to the English. The Persians would pay half of the supply costs of the fleet during the campaign. (In the event they paid only half of the agreed £600 per month.) The seamen were promised an extra month's pay, which won them over (Forster 1906: ix). Following victory, the English were to receive an equal share of the spoils and future customs duties at Hormuz, together with a special concession of trading in Iran free of taxes. The agreement included an additional clause which stipulated that prisoners would be dealt with according to their religion: Christians were to be surrendered to the English and Muslims to the Persians (Sykes 1915: 275–6). Such a degree of sensitivity on cultural and religious matters may be an indication that George Strachan was involved in the negotiations – and not just as an interpreter but as a counsellor. Muslim treatment of the Christian Portuguese would have been harsh – they executed most of their Muslim prisoners (Forster 1906: x) – and Strachan would not have wanted that weighing on his conscience.

There is no written record of George Strachan acting as interpreter for the East India Company. The absence of any record mentioning Strachan is, paradoxically, a clue to his being closely involved in their negotiations. Pietro Della Valle wrote in his journal that in October 1621 he set out from Shiraz intent on reaching Gombroon (now Bandar Abbas) to take ship to India (Della Valle 1664: vol. 2, 436–7). He and Strachan had previously agreed that they would journey to India together but, despite the fact that they had met in Isfahan the previous month (Chapter 11), the Scotsman did not travel with his party. Strachan must have been engaged in some serious matter. If he had been ill, his friend would have delayed his departure until he recovered.

In December at Minab, a small settlement thirty miles from Gombroon, Della Valle found his way barred because of the preparations for the Hormuz campaign. He met up with East India Company merchants who had stopped there to avoid risking their shipment of silks at the coast. The Portuguese had been sending raiding parties from Hormuz to the mainland to try to disrupt the preparations for the expected assault. The governor of Shiraz, Quli Khan, had allowed the English use of his official residence in Minab. When Della Valle made his presence in the town known, Robert Gifford, the Catholic member of the English party, whom the Roman described as 'a friend of mine of long standing', met with him and on behalf of Edward Monnox, who was indisposed, invited him and his party to stay with them (Della Valle 1664: vol. 2, 436–7; Bull 1989: 180–2). Strachan was not present in Minab with the English. Not only does Della Valle not mention the Scot but when his wife, Manni, fell fatally ill the Roman nobleman could not find a physician to attend to her. Strachan's services as a doctor would have been called on by his friend if he had been anywhere nearby. The explanation is likely to have been that he was with Quli Khan acting as liaison for Monnox and the East India Company. In one of his letters to the governors in London, Jeffris had described Strachan as 'the only councilor and director of the silly agent Signor Monox' (*Calendar of State Papers, Colonial, East Indies, 1617–1621*: 388). Edward Monnox had shown himself to be a resourceful and practical man and, since the success of the campaign required the coordination of the English and Persian forces, he would have had little hesitation in using his 'councilor' to liaise with the Persian commander. However, prudence would have caused him to commit nothing of this to writing. The following autumn Della Valle wrote that Strachan 'had all the interests of the nation [England] in his hands' as far as dealings with Persian authorities were concerned, and such a status would only be credible if he had participated in the Hormuz campaign as an agent of the English (Della Valle 1664: vol. 2, 224; Yule 1888: 319).

On 26 December a messenger came from the coast to Minab to tell Monnox that the English fleet had arrived in Jasques. He sent instructions that it should set off immediately for Kuhestak, a small harbour twenty-five miles east of Hormuz, where he was waiting for the ships when they arrived. The ships' captains told him that he had been dismissed from his post and that his replacement, Robert Jeffris, had died on the voyage from Surat. Monnox, through his strength of personality and knowledge of the English dealings with the Persians, continued to act as agent and direct the actions of the fleet against the Portuguese (Forster 1906: vii).

The East India Company provided five warships and four pinnaces, and at the beginning of January 1622 they launched their attack in accordance with the Persians' campaign plan. The Portuguese dominated the shipping channel of the Straits of Hormuz from their fort on Hormuz Island and another on the island of Kishm (now Qeshm) which is situated ten miles to the west. While the English ships blockaded the Portuguese fleet in its harbour at Hormuz, the Persians were free to land on Kishm and attack its fort. The Portuguese were unable to engage with the English fleet, part of which then sailed to Kishm to aid the Persian army. There they brought ashore their larger cannons and bombarded the fort. It was during this part of the campaign that the English suffered their most prominent casualty: Captain William Baffin, the Arctic explorer, was killed by a musket shot from the fort. In February the Portuguese surrendered, and the allied Persian and English forces were then free to lay siege to the fort on Hormuz. The action lasted three months, during which the Portuguese ships were destroyed at their moorings and the walls of the fort were breached (Sykes 1915: 277–9).

East India Company casualties were light and the company's share of the spoils amounted to £100,000. On hearing the news in England, the Duke of Buckingham was furious at the English involvement in the defeat of the Portuguese. His dealings with the Spanish monarchy regarding a marriage between Prince Charles and the Infanta were reaching a delicate stage. He threatened to withdraw the East India Company's charter. The governors were forced to placate him with a gift of £10,000 and shortly afterwards had to provide King James with an equal sum to satisfy him (Chaudhuri 1965: 64). The company's financial rewards from the campaign did not fully recompense the loss of three months' trading by their merchant fleet (Forster 1906: xii). The Persian promise of no taxation and a half share in duties paid at Hormuz did not materialise. They abandoned the island for trading purposes and greatly expanded the port of Gombroon located nearby on the mainland which they renamed Bandar Abbas in honour of the shah. Most of the Indian merchants who had traded on Hormuz moved to the new port, but a significant number relocated to trade at the Portuguese fort at Muscat (Onley 2014: 232) which had been reinforced by the survivors of the garrisons of Hormuz and Kishm on their release by the English. From their strategic bases at Muscat and Goa, the Portuguese remained a threat both militarily and commercially to English, Dutch and French merchants trading in the Arabian Sea, but they were no longer able to control the sea trade through the Persian Gulf.

Shah Abbas continued his military campaigns by taking Kandahar in Afghanistan from the Moghul emperor and conquering Baghdad in 1623. With total dominance of the silk trade, his empire enjoyed a short period of relative peace and prosperity. East India Company merchants benefited from the goodwill of the shah for their help in the capture of Kishm and Hormuz. However, this good fortune was limited and did not survive the death of the shah in 1629. Before this happened, George Strachan had ended his involvement with the company and again turned his attention to pursuing his travels eastwards.

CHAPTER ELEVEN

Among Friends

Catholic Religious Orders in Persia

George Strachan had left Baghdad in the spring of 1619 to travel to Isfahan with the intention of capitalising on the goodwill of the East India Company, which he felt he had earned by his actions in rescuing William Nellson from execution at the hands of the aga of Baghdad (Chapter 9). If he had failed to gain a positive reception from the merchants in Persia, his plan was to continue eastwards to India. As he explained in his letter to Sir Thomas Smyth, he was travelling to the court of the Great Moghul 'with good recommendations and fayre expectations' (Yule 1888: 324). He did not explain from whom he had been given recommendations or who had led him to believe that he would be well treated in India. It is possible that some of his Arab merchant friends had offered introductions to their trading contacts on the sub-continent, but it is more likely that members of Catholic religious orders were the source of his confidence in receiving a welcome at the Moghul court.

He had been reliant on the Franciscans while in Ottoman lands. During his stay in Aleppo, he spent time with them in the Convento di Terra Sancta. There were many Catholic missionary stations throughout the Middle East, India and the Far East. Prominent among the missionary orders was the Society of Jesus which was the most extensive in its reach, but when Strachan arrived in Isfahan there were no Jesuits in Persia. Jesuits and Augustinians had set up missions on the island of Hormuz in the late sixteenth century but suffered from health problems caused by its adverse climate. They had very limited success in making conversions, in part due to the fact that much of the population was non-resident. Merchants made brief stops during which they traded and amused themselves. Hormuz and Kishm were notorious for the abundant opportunities for drinking, gambling and fornication. As well as providing additional income for the residents, these facilities made the islands the trading station of choice for merchants (Coleridge 1997: 104–5). The missionaries attended to the Portuguese garrisons but after a few years both the Jesuits and Augustinians withdrew to Goa. In 1602 Portuguese Augustinians arrived in Isfahan and received permission from Shah Abbas to establish a mission. The shah provided them with a convent building and later they were able to set up a small priory in Shiraz. The Augustinians attempted to minister to Syriac Christians and although Georgian Orthodox priests objected to their presence, Abbas allowed them to stay. Any concerns he may have had regarding their activities stemmed more from their being Portuguese than being Christian but he exercised close control over them (Flannery 2013: 73–91).

In 1608 the Augustinians were joined in Isfahan by the Order of the Discalced (Barefoot) Carmelites. These friars had been recruited in Italy by Pope Pius V in 1604 specifically as emissaries to Shah Abbas to demonstrate the desire of the Christian West to cooperate with Safavid Iran against the Ottomans (Chick 1939). The pope had taken pains to ensure that no Portuguese were included in the group although he appointed a Spaniard to lead it. This was a sensitive matter in which Pius had to take care to offend neither the shah nor the Portuguese. Despite representing the combined monarchy of Portugal and Spain, Philipp II/III claimed that all missionary activity in Asia should be under Portuguese control in accordance with the Treaties of Tordesilhas and Zaragoza agreed in 1494 and 1529. The terms of the treaties, which had been drawn up with papal approval, allocated the territories being explored between the Portuguese and Spanish for commercial and religious exploitation. The treaties were disregarded by other European states especially those of the Protestant North but for a century, with the exception of the Spanish in the Philippines, Catholic missionaries largely complied with the artificial arrangement. The religious aspect of the treaties was ended with the establishment in 1622 of *Propaganda Fide*, the papal congregation that assumed worldwide responsibility for missionary activity.

The pope's Carmelite emissaries to Persia were led by Fr Jean-Thaddée de St-Elise (Fr Giovanni, as Della Valle called him) and, on their arrival in Isfahan, they were welcomed by Shah Abbas who installed the friars in a royal residence near the Meydān-e Mir, one of the great public squares in the centre of his new capital. This was a mark of honour and a sign of friendship but, as had been the case with the Portuguese Augustinians, it also allowed Abbas to have complete control of their movements and contacts. The Carmelites had been prepared for their mission in Rome by undergoing intensive language training in Farsi. As vicar general of the mission, Fr Giovanni engaged in discussions directly with Abbas and was able to develop a personal rapport with the shah. Abbas showed great curiosity in Western matters and in particular one of the friars' gifts – an illustrated copy of the Bible – appealed to him. He asked the friar to produce a translation of the Psalms in Farsi and instructed three mullahs and a rabbi to assist in the work. It was printed in the Carmelite convent and presented to the shah in 1616. Although the friars did not turn their possession of a printing press into a commercial enterprise, it was the first ever to operate in Persia (Gulbenkian 1981: 40).

The Carmelites are a missionary as well as contemplative order and, in addition to fulfilling their diplomatic embassy, they saw it as their duty to attempt to gain converts. The shah was tolerant of non-Islamic religions but, despite occasionally teasing European visitors with suggestions that he might convert to Christianity, he was a strong defender of Shi'a Islam (Bull 1989: 165–6). The Carmelites soon realised there was little chance of being allowed to make converts from among the Muslims and, like the Augustinians, they turned their attention to the 'third force' comprising the Christians of the Caucasus who had been forcibly relocated to Iran. Those who had converted to Islam were forbidden to apostatise, but those involved in the silk trade who, through Abbas' dispensation, had remained Orthodox Christian became targets of the Carmelites' missionary efforts. Initially these Christians welcomed the Carmelites, and some were persuaded to accept the authority of the pope. The friars set up schools for the Caucasian exiles and even conducted joint services with Georgian and Armenian Orthodox priests. Following these small successes, the Carmelites asked the shah to allow them to relocate their convent to the enclave of New Julfna. Already resenting the approaches of the Augustinians, the Orthodox metropolitan grew ever more alarmed at the encroachment of the Carmelites and appealed to Abbas to forbid the friars proselytising among his congregations. Abbas did not want his valuable silk industry disrupted and agreed. Although he allowed the Carmelites to move to a dedicated convent house, they were isolated in a Muslim part of the city with little contact with other Christians (Richard 1990: 832–4). From that point onwards they were restricted to ministering to visiting European Catholics and following a contemplative life.

Friendship with Pietro Della Valle

This was the situation that George Strachan found when he visited the Discalced Carmelites in 1619. He called on them before he sought out the East India Company merchants in the city. It is likely that it was through the friars that he was introduced to the Spanish ambassador, Don García de Silva y Figueroa, and later met Pietro Della Valle, the Roman nobleman whom he was to befriend. Strachan kept in regular contact with the religious order and when in 1620 he was dismissed from the service of the East India Company he took refuge in their convent, being dependent to an extent on their charity for several months (Chapter 10). Before those events happened, he and Della Valle had already become close friends. The first mention that the Roman makes regarding meeting Strachan is in a letter dated 24 August 1619, only two months after the Scotsman had first met the merchants of the East India Company. It is clear that even at that early date, the two men had spent time getting to know each other. Della Valle had heard of Strachan while travelling across the Great Syrian Desert and knew of his time in the household of Emir Feyyād. Earlier in 1614, although Strachan had left before Della Valle arrived, they had both been guests of de Harley, the French ambassador in Constantinople, but they had more things in common than their mutual host. In his journal the Roman nobleman explained why they had become close friends:

> For since the first day of our acquaintance, through a congeniality of spirit, and a conformity of ways, in addition to an equal delight in study, and that zeal and love for our common faith, which in these regions has served to make a stronger tie between us than anything else, there has arisen between him and me a most intimate and unbroken friendship. (Della Valle 1664: vol. 2, 718; Yule 1888: 318)

Della Valle soon realised that Strachan was the better language scholar. Their first discussions centred on their collections of books and Strachan agreed to undertake a major work suggested by Della Valle:

> He has promised me that he will apply himself to the translation of the Camus, which is the most ample and perfect dictionary that the Arabs have, and I have myself a first-rate copy of it, which should be known in Rome. If Signor Strachan should undertake this task, the work will be worthy of passing through the hands of scholars. (Della Valle 1664: vol. 2, 718; Yule 1888: 316)

The work referred to is *Al-Qamus al-Muhit* by Al-Firuzbadi, the early-fifteenth-century Iranian lexicographer. It is, even today, a major authoritative dictionary of Arabic and contains over 60,000 entries (Nahli et al. 2016). Its translation would have taken years to complete and there is no record of Strachan even starting the translation; certainly he did not complete the task.

Despite their genuine mutual interest in Eastern scholarship, the two men seem to have valued their friendship largely for social reasons. The Roman was missing his homeland and found the Scotsman's company extremely congenial (Yule 1888: 315). Strachan, on the other hand, would have welcomed his conversations with a social equal as relief from his dealings with the English merchants and their preoccupation with trade. The two friends were able to converse in Italian and, given Strachan's familiarity with Rome and his many dealings with the Church's hierarchy, they would have had acquaintances in common. Della Valle belonged to an old aristocratic Roman family which included two bishops and a cardinal (Rossi 1953: 49–64). In less than a year, Strachan's friendship with Della Valle was strong enough to extend to members of his new eastern family. He was asked by Maani's brother, Abdullah Giorido (Della Valle's Italianate spelling of his Georgian name), to be godfather to his newborn son. The infant was named George, possibly in honour of Strachan. The baptism was conducted in Isfahan on 3 February 1620 by the Carmelite vicar general, Fr Giovanni, but Strachan was indisposed and Robert Gifford stood proxy. This may have been a forerunner of the serious attack of malaria that Strachan suffered in Shiraz three months later, or possibly he was acting as interpreter on Monnox's behalf in his dealings with Quli Khan (Della Valle 1664: vol. 2, 157–8).

Despite their close relationship, Strachan never confided in Della Valle on the reality of his dealings with the East India Company merchants. The Roman's understanding of the relationship was that he was a guest of the English:

> Becoming known to the English as a gentleman of their nation, and one of such eminent capacity, although by open profession a Catholic, they insisted on having him in their house, where they lodged him and continued to entertain him in the most honourable manner. And he always stopped with them, except once for a while, a little before I left Ispahan, when, for what reason I know not, he put up for some months at the convent of our Barefoot Carmelites . . . He went back to the English, nor do I know the reason for that either. (Della Valle 1664: vol. 2, 720–1; Yule 1888: 318)

Through pride or personal vanity, Strachan did not tell his noble friend that he was a salaried employee or that he had been dismissed because of the malicious gossip of Robert Jeffris. Della Valle was fully aware of his service with Emir Feyyād, but as his medical adviser the emir treated and rewarded the Scotsman in keeping with his profession. Despite the problems it brought, being accepted into the emir's family was a clear recognition by his host of Strachan's noble family background. His treatment by the East India Company, in which he was officially a company servant, could not be viewed in the same light. Any sense of shame regarding his status within and treatment by the East India Company would have been enough to prevent him being completely frank with Della Valle. For the same reason, Strachan would not have discussed any part that he played in the negotiations between the English and the Persians regarding preparations for the attack on Hormuz. He would have known how his friend would have viewed such involvement. Della Valle expressed his feelings at the time:

> with intense distaste for the torments I saw being prepared for the Portuguese Catholics, whose hurt (as a nation deserving so well of the Church and of God, and so glorious among the rest of us Christians for its worthy enterprises) I felt in the extreme. (Bull 1989: 183)

Della Valle knew that Strachan's stay in Persia was temporary and that his plan was to travel to the court of the Great Moghul in Agra (Dellavida 1956: 65). He also appreciated that, as a younger son of an impoverished nobleman, Strachan was not wealthy and had suggested that he should join his party and travel with them to India as his guest. Della Valle remained in Persia for a further three years but when he did leave Strachan did not accompany him. However, during this period the Scotsman continued to plan his onward journey.

Safeguarding the Library

In May 1621 a small party of Carmelite friars arrived in Isfahan from Rome. It included Fr Vincent of St Francis, who had been sent as visitor general to carry out an inspection of the Carmelite missions in Persia and India, and Fr Prospero of the Holy Spirit, who had come with him to take up the position of prior of their house in Isfahan. When they arrived, they found George Strachan resident in the monastery teaching Arabic to the friars. He remained with them until September when the prior reported in a letter to Rome that:

> Although we have not the funds for our food I did not wish to miss the opportunity of [profiting by] a man eminent in Hebrew, Greek, Chaldean and Arabic . . . the fathers will assist him to make his journey [to India] by paying him 50 scudi. (Chick 1939: 236)

The sum of fifty scudi was inadequate for such a venture, but Strachan had not ended his working relationship with the English merchants and was able to save more. As well as the small amount of money, he benefited in another way from his association with the Carmelites. Fr Vincent's intention of travelling on to India overland had been thwarted due to the Persian preparations for war with the Portuguese. Border crossings were closed and travel eastwards was forbidden. He had no means of carrying out his inspections in India and was preparing to return to Rome. Strachan saw this as an opportunity to safeguard his precious collection of books. He clearly ran risks in transporting the books on a journey east, but he may also have been concerned for his personal safety given the prospect of the imminent outbreak of war with the Portuguese. Although not a combatant, as interpreter and intermediary, he would have been close to the fighting. He decided to entrust his library to the visitor general to take with him to Rome. In September 1621 two legal documents were drawn up to validate the arrangement. In the first Fr Vincent declared that he had received from:

> Mr George Strachan, a Scottish gentleman, a coffer, in which there are books in Arabic, Persian and Turkish, some 61 in number, listed according to a catalogue, together with all the books which he [Fr Vincent] was to recover in the city of Aleppo there lying in the name of Mr George Strachan, and promised to take them with him to Rome, expenses being paid by the Carmelite Order, and that he will guard and keep them safe and sound in the Convent of S. Paul of the Discalced Carmelites until the arrival of the said Mr George Strachan and then, as soon as Mr George Strachan reached the city, would return them to him, when the expenses incurred in their transport had been repaid to the Carmelite Order: all this without any responsibility and, if anything should happen to the books, (which God forbid) he, Fr Vincent was not bound to pay any compensation. (Chick 1939: 235)

The second document was Strachan's will:

> I, George Strachan, Scot of the Mearns, declare that this is my will. Since we are all mortal, should God dispose of me during the journey which I undertake, I spontaneously and freely give and piously bequeath as alms all the aforesaid books to the Monastery of the Saint Paul Mission of the Discalced Carmelites in Rome, so that the Fathers and Monks of the monastery may pray for my soul and the atonement of my sins in their masses, orisons and penitences. Such is my last will. Made in Ispahan on the 16th of September in the year of our Lord 1621. (Chick 1939: 235; Dellavida 1956: 65)

Della Valle was witness to the will, which demonstrates the closeness of the bond that had formed between the two men. They had agreed to travel to India together and had decided that Goa would be their first stop. This choice appears to have been largely at Strachan's suggestion but, when Della Valle decided to move on in October 1621, Strachan did not accompany him. Following Strachan's involvement in the attack on Hormuz, it would have been dangerous for him to travel to Portuguese territory. This is likely to have been the reason that Strachan could not accompany his Roman friend and would explain why he was unable to discuss the reasons for his change of plan. It was during the delay caused by the impending hostilities that Della Valle lost his wife, Manni, while they were staying with the East India Company merchants in Minab. By his own account, he went into a deep depression and even contemplated suicide (Della Valle 1664: vol. 2, 439). His party abandoned the journey to the coast and returned to Isfahan where he gradually recovered his spirits, but it was nearly a year before he resumed his travels. By then Della Valle was aware that Strachan had returned to working for the English merchants, and in his diary he provides the last confirmed written record there is of the Scot's life.

Farewell

Della Valle's grief over the death of his young wife was sincere. In Isfahan he had her body embalmed with the intention of taking it home with him to Rome to be buried in the family vaults. He wrote that he needed to be circumspect in his actions in order to achieve this. His return required a number of sea voyages and such was the superstitious attitude of sailors that they would have refused him passage, if it were known that he was travelling with a corpse. Manni's remains were concealed inside an expensive Persian carpet and conveyed to the coast as part of the collection of souvenirs of the Roman nobleman's peregrinations. These mementos included two Egyptian mummies that he had collected in Cairo which also had to be disguised. After making his elaborate preparations, he and his party set out for Gombroon to take ship, as he hoped, for Muscat on the initial stage of his journey to Goa.

He had been delayed in the port waiting for permission to leave when George Strachan arrived on 24 October 1622. He had come in advance of a caravan of English merchandise with the task of renting a house in the port, which could be used as a storage depot and trading post for the English. Strachan called on his friend and brought news from Isfahan with a letter that he had been asked to deliver by the Carmelite prior, Fr Prospero. The Scot had arranged to meet the military governor of Gombroon, Sevenduk Khan, the following day, and Della Valle asked to accompany him in order to plead his case for permission to sail from the port. He had been waiting for a decision from the governor, whom he described as the sultan, and had been unable to obtain an audience while he waited. Strachan may have had some misgivings about the Roman being present, since the object of the meeting was to discuss the English merchants' trade privileges promised as reward for their participation in the capture of Hormuz, but he agreed (Della Valle 1664: vol. 2, 722–3).

Della Valle's account of the meeting is informative of the respect that Strachan was shown by the Persian khan. The Roman addressed the governor, saying that he could be trusted because he was a friend of the English. Strachan confirmed this and spoke warmly of his friend. Della Valle wrote that, at this point, the sultan spoke openly to Strachan in a relaxed manner that he had not used with himself. The sultan explained that he had written to Quli Khan in Shiraz asking for instructions on how Della Valle should be treated 'in such times of suspicion' – should he be allowed to proceed or be detained? (Yule 1888: 319). He was still waiting for a reply from the provincial governor but, given Strachan's assurances, he was prepared to let Della Valle leave for Muscat on the next available ship. Since that would not be earlier than in twenty days' time, he expected instructions from Shiraz to arrive before then, and he would act on them. The Roman left the Scot alone to conduct his business with the sultan and on his friend's return to his lodgings, Strachan explained that the sultan had strongly advised against Della Valle travelling to Arabia. He had warned that, because of the recent hostilities, the Roman was likely to be viewed with suspicion by the Portuguese authorities in Muscat and not be allowed to proceed. The friends discussed this advice and Della Valle decided to travel on an English ship to Surat instead. Strachan assured him that, whatever instructions regarding him arrived from Shiraz, he would use the powerful influence that the English had with the shah to ensure that neither Sevenduk nor Quli Khan could prevent his leaving (Della Valle 1664: vol. 2, 724). This may have been unwarranted braggadocio on Strachan's part, but it left his friend with the clear impression that the Scotsman 'at this time had all the interests of the nation in his hands'. As far as the Persian governor was concerned, Strachan spoke for the English. The regard in which the sultan held him is likely to have been the result of Strachan's intimate involvement in the successful outcome of the Persian-English alliance against the Portuguese.

The two friends remained in Gombroon for another month: Della Valle waiting for his ship to Surat and Strachan handling the English merchants' affairs after the arrival of the first caravan of goods under the supervision of Captain John Benthall on 28 October. A month later, on 29 November, Della Valle wrote that Strachan had fallen seriously ill with a fever. It is likely that this was a recurrence of malaria, as it was decided that he should leave immediately for Lar, the nearest city in the mountains, where the air was better and he would be more comfortable than in his lodgings in the port. Della Valle entrusted the letter in which he was describing these events to his friend in the belief that the Scotsman would recover fully, and afterwards be able to travel to Isfahan and arrange for it to be forwarded to Italy. Strachan left that same night and the two friends never met again (Yule 1888: 319).

There is no record of Strachan's movements following his last meeting with Della Valle. There has been speculation on his eventual fate but there is little direct evidence after 1622. In his biography of Strachan, Thomas Dempster asserted that his friend was living in Persia having been travelling in the East for nine years. This would indicate that he last heard from Strachan in 1622, possibly in a letter sent along with Della Valle's from Isfahan that year and carried by Fr Vincent. Dempster died in Padua in 1625 and does not appear to have received any further information from Strachan. On his return to Rome, Della Valle tried to keep informed of his friend's actions. He was aware that he had survived the bout of malaria he was suffering from when they parted since the letter he had entrusted to him had been delivered. In a letter dated 26 October 1630 written to Sebastian Tengnagel, the custodian of the imperial library in Vienna, he mentioned Strachan's collection of manuscripts kept by the Discalced Carmelites in Rome, referring to the Scot as 'once an intimate friend of mine, who died some time ago on his way back from India to Persia' (Dellavida 1956: 69). The Roman appears to be wrong in this regard. There is convincing evidence that Strachan was still alive in 1634. The mention of Strachan having visited India, however, seems to be based on fact.

An Independent Trader?

In early 1623, when Della Valle sailed from Gombroon to Surat on *The Whale*, the East India Company ship (Bull 1989: 193), he had every reason to believe that the fortunes of the English in Persia were set fair for prosperous trading. However, he had witnessed the high-water mark of their success and, in a relatively short time, the company's circumstances underwent a significant change for the worse. Trading in Persia was always less advantageous to Europeans than in India or the Spice Islands. Iran was a relatively poor country with few natural resources and a small population – fewer than eight million during the reign of Shah Abbas (Matthee 2018: 50). The only significant exports of value were the high-quality finished silk fabrics and carpets. To a large extent, the shahs of the Safavid dynasty relied on trade to support their rule and, because of this, controlled as firmly as possible the wealth derived from it. Abbas was more successful than his predecessors. He closely guarded the collection of taxes and grew the revenues by promoting state manufactures and encouraging self-sufficiency. He had expanded his empire and reinforced his authority through military conquest and strategic alliances, entering into treaties with foreign powers at critical times to protect his own interests. The terms of the humiliating Treaty of Constantinople that he had been forced to sign in 1590 were broken as soon as he was able to confront the Ottomans militarily. The peace treaty of Nasuh Pasha which ended the war in 1612 included a clause that tribute would be paid by Persia to the Sublime Porte, but Abbas ignored that provision, thus provoking the renewed hostilities of 1618. In his disregard of treaties when it suited his interests, Shah Abbas was following the Shi'a doctrine of *taqiyya*. As a persecuted minority for much of their history, adherents of Shi'a Islam had rationalised dissimulation as a means of survival. This doctrine allows the making of promises which can be broken when necessary.

The shah's agreement with the English was drawn up to obtain the use of their fleet. Once Kishm and Hormuz had been captured, Abbas had little need for the English. While it was not in his interest to antagonise them unduly, the lavish trading concessions which had been agreed were no longer necessary. In his eyes, the English had been rewarded sufficiently by their share of the spoils. The promised half share of import and export duties through Hormuz was never going to materialise. It became the first element in the shah's reneging on his agreement (Lockhart 1958: 361–3). The decision to enlarge the port of Gombroon, renamed Bandar Abbas, as the replacement for Hormuz and Kishm made eminent sense to Abbas. With no viable navy to defend the islands, it was in Iran's interest to have trading conducted at the mainland port. The English merchants were required to renegotiate terms and had to accept Abbas' offer of one third share of duties instead of the previously agreed half (Matthee 2018: 44). Abbas may have made this new offer with the thought that he might have need of the English fleet in the future, but he did not honour even the reduced commitment. They received no share of the port duties and their initial exemption from paying taxes on their own imports and exports was soon removed. The East India Company merchants were to conduct business in Persia on similar terms to those of other European traders.

In 1623, the year following the victory over the Portuguese, the reputation of the English as a major force in the East took a severe blow. The Massacre at Amboyna, as it became known, saw the Dutch forces of the VOC execute all of the English merchants at the trading station at Amboyna in the East Indies. The English account of the incident (Skinner 1624) claimed that nearly one hundred servants of the East India Company, along with a small number of Japanese traders, were tortured and publicly beheaded in order to deter competition and intimidate the local rulers who were seeking better prices for their spices than the Dutch offered (Ryu 2009: 178–89). The elimination of competitors solidified Dutch dominance of the spice trade and, although it caused an outcry in England, neither King James nor the United Provinces wanted it to damage their cooperation in the Thirty Years' War. Apart from making protests, the English took no action in retaliation. The lasting outcome was that the East India Company withdrew entirely from the East Indies and Japan and restricted spice trading to India. That same year, 1623, Shah Abbas marched his army into Mesopotamia and captured Baghdad without any involvement of the English. George Strachan's boast to Pietro Della Valle that the English had powerful influence with the shah may have been true when he made it, but it was short-lived. In the East the East India Company was not to be a dominant force in either politics or trade.

Strachan's work for the East India Company as interpreter and intermediary – in effect the company's dragoman – did not last long. In a letter dated 24 January 1623, William Bell, the replacement agent for the East India Company in Persia, wrote in reply to a query from Surat that 'Mr Strahand is longe sence dismissed the Companies service'. In the same letter he reported that Mr Monnox was returning to England, much to Mr Bell's relief, and that Mr Cardo, the minister, had died before he could be sent home as instructed (Forster 1906: 24). In Strachan's case, Bell was exaggerating if not lying. Della Valle's account proves that Strachan had been employed by the company only two months earlier. For a year Bell had been company agent in name only; Monnox ran the factory in defiance of the instructions from London. Bell was unwilling to admit this, and his relief at Monnox's departure for England is understandable. No matter how useful the Scotsman could be to him, Bell would not have had the courage to continue to employ him.

Strachan continued to live in the East for more than a decade after his employment with the East India Company ended. This could only have been a deliberate choice on his part. A journey back to Europe by sea had been offered by Pietro Della Valle, and Edward Monnox may have been amenable to Strachan's accompanying him home. In addition it would have been possible for him to return overland since Emir Feyyād was no longer in control of the desert. When news of King James' death in 1625 came, he may have thought that a return to Scotland was at last open to him. It is unlikely, however, that he would have heard any news from his family. King Charles had favoured Strachan's nephew, Sir Alexander, by appointing him a Commissioner of the Exchequer and awarding him a baronetcy of Nova Scotia. This good fortune meant that the family's financial difficulties were at an end, although, by acting as the king's tax collector, Alexander gave himself problems regarding the nobility and gentry as well as the civic authorities. It is probably fortunate that George was unaware of this since his return to Scotland would not have gone unnoticed or uncontested. He stayed in the East because of his desire to continue with the life he had created there for himself.

It is likely that he remained in contact with the Carmelite priors but would not have been dependent on their charity. The fathers were too poor to have provided him with more than a basic subsistence. Continuing his studies in Eastern literature and fulfilling his ambition to visit Agra and the court of the Great Moghul would have required a significant income. He must have worked as a physician or taught languages and it is possible also that he engaged in trade. His language skills and familiarity with Arab and Persian society had proved extremely useful to individual English merchants in their private trading. Some of them may have been prepared to defy Bell and the company to act in partnership with Strachan. This would have allowed him to travel between Persia and Surat using the company ships, and an unknown report of such activity may have been the source of Della Valle's belief that Strachan had died while returning to Persia from India. He may also have traded in partnership with his Arab merchant friends who operated out of Aleppo and Baghdad (Yule 1888: 325). Arab and Indian traders sailed out of Basra as well as the Persian and Portuguese controlled ports of the Gulf. By engaging in their enterprises, Strachan could have visited ports in India including Goa where his personal contacts could provide him with an introduction to the Moghul court in Agra (Chapter 12). This is largely speculative, however. Much of the available evidence suggests that he spent the greater part of his later life in Persia. While there he would have been restricted to earning a living as a physician or tutor probably servicing the resident European community.

Shah Abbas' Divan and Mīr Dāmād

Strachan had boasted to Della Valle that, if Quli Khan instructed Sevenduk Khan to refuse the Roman's request to sail from Gombroon, he would use influence at the shah's court to ensure his friend could leave as he wished. Since the only person who could overrule the provincial governor was Shah Abbas, Strachan must have felt he had very friendly relations with the shah and his court. This is not as impossible as it might appear. Abbas is known to have involved himself in every major aspect of his empire's affairs and would have played a personal role in the negotiations with the English. He was also extremely inquisitive about foreigners and would have taken time to interrogate the Scotsman (Bull 1989: 163–70). The two men were of a similar age and Abbas could only have been impressed with the Western scholar who was fluent in Arabic and Persian and knowledgeable about his country's literature, especially poetry. Abbas filled his court with literati, and Strachan would have been delighted to have been in such a cultured and aristocratic company as was represented in the shah's divan.

The Persian State archives may contain references to Strachan's presence at court. There has yet to be significant work done to determine whether such material exists. (It is to this author's great regret that he has neither the language skills nor opportunity to research the archives of Shah Abbas.) However, recent scholarship in unrelated areas has uncovered two references to Strachan which may place him at the court of Shah Abbas. The first reference occurs in research conducted into Arabic studies in mathematics by Gregg De Young (De Young 2015). Reference has been made to Strachan's copy of Euclid's *Elements* that De Young discovered in the Majlis Shūrāh Library in Tehran (Appendix: uncatalogued Ms. No. 4 of 6). Although undated, Strachan must have bought it after Fr Vincent left for Rome in 1622.

The second reference concerns another manuscript which was found in Baghdad in the 1810s by Claude Rich, an East India Company official based in Persia (Appendix: uncatalogued Ms. No. 5 of 6). The bound manuscript contains two books in Persian each of which has a Latin interlinear translation by Strachan. The first is a treatise on logic by a fourteenth-century Persian author, Alī ibn Muḥammad al-Jurjānī. The second book, *Jām-i gītī-numā*, is more important for the light it throws on Strachan's life. It is a book on popular philosophy by Mīr Ḥusayn al-Maybudī and has, written in Strachan's hand on the flyleaf of the manuscript, the comment 'translated into Latin by George Strachan, Scot of the Mearns, 1634'. Since the date is crucial, Professor Dellavida had the manuscript checked to ensure that there was no error in reading the inscription (Dellavida 1956: 71). This proves that Strachan was still alive and well in the East four years after Della Valle believed him to be dead.

The book provides more information on the Scot thanks to recent work by Reza Pourjavady. In the course of his research into works of Persian philosophy, he has discovered that Strachan's glosses on *Jām-i gītī-numā* show that he was familiar with the philosophical arguments of the leading scholar of Isfahan at the time, Mīr Dāmād (Pourjavady 2017: 565). Mīr Dāmād was a mathematician and poet as well as a philosopher who was a prominent member of the imperial court. He was also Abbas' son-in-law. The shah commissioned him to produce the intricate geometric designs on the tile cladding of some of the impressive new mosques in Isfahan. At court, Mīr Dāmād gathered a circle of scholars around him to engage in learned discussion (Nasr 2006: 214). Such a grouping would have been irresistible to Strachan, and he must have tried to join it. Spending time in learned company and enjoying stimulating conversation was important to him. (It was for such a reason that he had delayed his departure from Emir Feyyād to continue discussions with his Islamic teachers.) Without further evidence, it is impossible to say whether Strachan was part of Mīr Dāmād's circle or had simply read his work. Membership of this elite group would have been enough to keep him from returning to Europe.

When he translated *Jām-i gītī-numā* in 1634, George Strachan was in his early sixties. His life had been that of a wandering scholar and if, as he wrote, he was intent on visiting Agra he would not have left it so late in life. It would be more in keeping with the nature of the man that at some time during the twelve-year lacuna in our knowledge of his travels that he found the opportunity to go to Agra. It appears, however, that the 'good recommendations and fayre expectations' that he boasted of in his letter to Sir Thomas Smyth did not materialise and, because of *Jām-i gītī-numā,* we know that he had returned to Persia by 1634. Any account of Strachan's life after 1634 would be entirely speculative if it were not for one other piece of evidence, appropriately a book, which sheds light on how Strachan appears to have spent his last days and how his life ended.

CHAPTER TWELVE

The Mission at Srinagar

Jesuit Arrival in India

After returning to Rome in January 1604, Strachan met with Claudio Aquaviva (Chapter 3). At considerable cost to himself, he had fulfilled his commission to deliver the Jesuit general's letter to Fr Robert Abercrombie. Divorced from his family and permanently exiled from his homeland, Strachan was at a critical point in his career. His discussions with Aquaviva must have touched on his future but, even although he was by then in his early thirties, he remained undecided about becoming a Jesuit. Abercrombie had informed the general that Strachan wanted to enter the Society, and Aquaviva may have tried to persuade the young man with accounts of the heroic challenges offered by the Jesuit mission. Twenty years earlier his nephew, Rodolfo Aquaviva, had led the first mission to the court of the Great Moghul, Akbar, and, shortly afterwards, along with four Jesuit companions had been murdered by Hindus. This event may not have featured in the two men's conversation but Strachan would have been fully aware of the martyrdom. The details had been widely circulated in publications throughout Catholic Europe (Hebermann et al. 1913: 'Martyrs of Cuncolim'). It may have been the recollection of this conversation that caused Strachan to mention in his letter to the governor of the East India Company that he was travelling to the court of the Great Moghul 'with good recommendations and faire expectations'. Given his earlier dealings with the Society at the highest level, he would have felt sure of being welcomed by the Jesuits in Goa, and through them of being introduced to the Moghul court.

Jesuits were making their greatest contribution to the work of the Church through higher education but, when Ignatius Loyola and his companions first banded together as a society in 1539, their presentation to Pope Paul III for recognition was to be sent as missionaries to the East, particularly the Holy Land. The pope did not see this as a priority. Eastern Christians, for the most part, were able to survive and even flourish under Ottoman rule, although they were subjected to heavier taxation than Muslims. The aga of Jerusalem tolerated both Christian pilgrims and the Franciscan friars who ministered to them from their convent of St Saviour, for the excellent reason that they were the city's principal source of income. For the pope, a better use for the highly educated Jesuit order was to counter the arguments of the Protestant reformers. When the Council of Trent recognised the validity of the reformers' arguments regarding the quality of the education provided to clergy, Pope Paul directed Loyola and his Society of Jesus to rectify the deficiency by running the colleges needed for the training of aspirants to the priesthood. Fulfilling this directive became their principal occupation in Europe, but they did not abandon their ambition of engaging in missionary work in the East.

One of Loyola's closest followers, Francis Xavier, acting on a commission from King João III of Portugal, went to Goa in 1542 to minister to the Portuguese colonists who were abandoning Christianity and adopting Indian wives and customs. Pope Paul supported his mission and appointed him apostolic nuncio in the East. Although he was the first Jesuit to start missionary work in India, he was not the first priest. Goa had already been established as a diocese of the Church in 1534 and, even before the Europeans arrived, there were Christians of the Church of the Apostle Thomas, which is believed to have been founded in Southern India in the first century. Xavier found in Goa that Franciscans and secular priests had established the small College of St Paul for the purpose of training priests from among their converts. With the authority of both the king and the pope behind him, Xavier used St Paul's College as his base and began educating the children of the colonists and local Hindu converts. The college was successful and grew such that, by the late sixteenth century, it had over 3,000 students (Costa 2006). Xavier's actions in regard to education together with his preaching brought the 'errant' Portuguese back to the Church (Ames 2012: 12–15).

Having achieved his first goal, in 1546 he left Goa for the Far East. Before then, he wrote to the Portuguese king asking that the Inquisition be brought to Goa. Pope Paul had established the 'Congregation of the Holy Office of the Inquisition' with his papal bull, *Licit ab initio,* in the same year that Xavier had set out on his mission to the East. Paul had done so for the same reason that he had given the Jesuits permission to form their society – to counter the 'heresy' of the Protestant reformers. In making his request, Xavier expressed his concern that the majority of Portuguese merchants in the colony were 'New Christians' (Jews forced to convert) whose conversion was suspect. He saw the Inquisition as a means of testing the faith of the 'New Christians', as well as reinforcing that of Indian converts (Rao 1963: 43). Xavier was dead before the Inquisition arrived in Goa, but his vision was that the whole of the Portuguese East, from the east coast of Africa to China, should be treated as one missionary entity with Goa as its centre. Other Jesuits soon followed him, and St Paul's College in Goa became their headquarters in the East.

The Goan Inquisition

Jesuit successes in religious conversions did not disguise the difficulty that the social structure in India presented. Much recent scholarship has dwelt on the degree to which European concepts of Indian society and beliefs are artefacts of colonialism, particularly in reference to Hinduism and the caste system (Pennington 2005). The degree to which the disparate gods and beliefs of India coalesced into a single religion, 'Hinduism', by dint of colonial domination is being argued, but it is generally accepted that the process by which this happened started before European Christian intervention. Moghul conquest and contact with Islam changed Indian society, creating complex social relations (Lorenzen 2005). The caste system, as the social stratification represented in present-day India, did not exist in the same way when Francis Xavier preached among the Paravars in 1543. The Paravars were a community of fisher folk especially noted for pearl diving, and can be viewed best as representing an occupational caste. Xavier recorded that although the general population responded positively to the Christian message, the Brahmins (the priestly elite) did not. This was due to their perception of it being incompatible with their status of superiority. The Portuguese had no difficulty accepting social stratification since their society had similar characteristics, but Hindu concepts of uncleanness meant that Christian priests who ministered to the *Paṛaiyār* (pariahs or untouchables) were themselves contaminated, and should not approach Hindus of higher status (Aranha 2014: 215). This violated European concepts of the sanctity of priesthood and was to become a major difficulty.

When Xavier first arrived, there was relative toleration between religious groupings but that changed dramatically under the influence of the Office of the Inquisition. In 1560 by order of the Medici pope, Pius IV, the Inquisition was introduced to Goa. The Goan Inquisition, in the running of which the Jesuits played a major part, became one of the most feared and repressive manifestations of the Holy Office in the whole Church. It was driven by religious fanaticism and greed. Few were safe from its subjections. During the two centuries of its existence, more than 16,000 people were arraigned before it, although fewer than forty were executed (Ames 2012: 12–15). Those found guilty of heresy, as well as being physically punished – the extreme form being burnt alive at the stake – were stripped of their property with half the value being given to their accusers and half to the Church. Under these circumstances, no one could feel secure, but the greatest number of the Inquisition's victims came from the Hindu community. In their ambition to make the Portuguese colonies Christian, a major part of the Holy Office's efforts was concentrated on destroying Hindu culture, even to the extent that the local language, Konkani, was outlawed (Dellon 1687).

George Strachan had declared that it was his ambition to visit the Moghul court at Agra. If the 'good recommendations' he wrote of came from the Jesuits, his first port of call in India would have been their headquarters in Goa. His visit could only have been in the decade between 1624 and 1634. This was a period when the Inquisition was at its most fearsome. Although he may have believed that his previous connections to the Society of Jesus would have guaranteed him a welcome, it must have become clear to him while there that the activities of the Inquisition made Goa a dangerous place for foreigners. He would not have been immune from risk. The Portuguese would have known of his association with the East India Company and also of his time in the desert with Emir Feyyād. Their suspicions would have been aroused as to possible political reasons for his visits to Goa and the sincerity of his Christianity. His defence against accusations regarding his faith would have come from the Carmelites in Isfahan who knew him well and could vouch for his Catholicism. A greater cause for Portuguese concern would have been his association with the English. Being a Scottish Catholic would have helped him greatly in this respect. In Europe, Scots exiles benefited from the general admiration for Mary, Queen of Scots, whose execution was considered a religious martyrdom as well as a political murder. These sympathies had scarcely diminished in the four decades since her death. In 1627 the Spanish poet Lope de Vega published his hugely successful verse epic *La Corona Tragica* which depicted the queen's life and death. It had been dedicated to Pope Urban VIII and Jesuits would have read the poem. However, Strachan's greatest protection against scrutiny by the Inquisition was that he was not a wealthy man and did not excite envy. There would have been little financial reward in accusing him of heresy and, as is usually the case with informers, they would have run the risk of making enemies. With no obvious reason for arraigning him before the Inquisition, he would have been relatively safe, but, being an intelligent man, Strachan would have not remained long in Goa and would have been cautious while there.

Jesuits at the Moghul Court

In the latter half of the 1620s, when Strachan is likely to have made his first visit to St Paul's College, the Jesuit presence at the Moghul court was in decline. The first mission to Akbar's divan in 1580 was in response to a firman sent by the emperor to Goa. The Jesuit group that responded was led by Rudolfo Aquaviva. They came to the conclusion that the Great Moghul was only intellectually curious about religious matters, and despite his friendship he had no intention of converting to Christianity. After three years they withdrew, but Akbar continued to send invitations to Goa for Jesuit priests to attend his court. In 1591 a second mission was sent, but took only one year to come to the same conclusion as the first. The third mission went to the court in Lahore in 1595 and, although Akbar remained friendly, he did not change his outward adherence to Sunni Islam. It was during this visit that he gave the Jesuits permission to set up a college in Agra and churches both there and in Lahore. When he died in 1605, his son Jahangir continued his benevolent approach but this began to change after his death in 1627 (Maclagan 1932: 23–51).

Due to their commercial activities in Moghul territory in Bengal, Shah

Jahan, Jahangir's son, distrusted the Portuguese. In 1631 this led to open warfare. Jahan had a strong antipathy towards the Jesuits which went beyond any commercial rivalries. He saw their missionary work as disruptive of the religious harmony he sought between his Muslim and Hindu subjects (Ikram 1964: 175–88). His eldest son and heir apparent, Dara Shukoh, did value their erudition, however, and in 1656, when his father was in his dotage, he engaged a Jesuit, Fr Stanislaus Malpichi, as tutor (Wessels 1924: 88). This warming of attitudes ended in 1658, when Dara was defeated in battle by his younger brother, Aurangzeb, who deposed their father and ascended the throne.

No invitations were issued to the Jesuits of Goa to attend Shah Jahan's court or engage in debates. Despite changes in the political sentiment, the Jesuit college and churches in Agra survived under Moghul protection. If Strachan had visited Agra during Jahangir's reign, it is highly likely that he would have received a welcome at court due to his proficiency in Persian, the language of the court, and his interest in and knowledge of its poetry, which Jahangir and his courtiers were known to enjoy. Any visit later than 1627 would not have resulted in his receiving an invitation. There is no evidence that he attended the court, but due to his natural curiosity and long-stated intention to go to Agra it is highly likely that he did visit the city and it would have been in keeping with his previous behaviour to have lodged at the Jesuit college.

Jesuit Exploration of Central Asia

During any stay in Goa or Agra, Strachan would have become familiar with the Jesuit activities in the East. As part of their missionary work they engaged in exploration and reported their discoveries in publications that were circulated widely in Europe. Their expeditions were motivated in part by a desire to find the mythical Prester John of medieval fame. The existence of a Christian king in the East had been rumoured since the receipt of a letter supposedly sent by him to the Byzantine emperor, Manuel I Komnenos, in 1165 (Kurt 2013: 1–2). The letter offered military help against the Muslim forces attacking the Crusader kingdoms of the Middle East. The story was a fabrication born out of a pressing need to provide the Christian defenders of the Holy Land with hope of external help. No help materialised from that source, but in Europe the legend of the great Christian kingdom in the East refused to die. Over time the fable of Prester John became conflated with the Christian empire of Abyssinia, whose pilgrims visited Jerusalem. The original legend described his kingdom as being somewhere in Asia beyond Persia in the land of 'the three Indies' (Kurt 2013: 2–3). Abyssinia and its Ethiopian empire did not fit this description, and by the beginning of the sixteenth century, when the Portuguese had gained greater knowledge of the emperor of Abyssinia, it was determined that Ethiopia could not be the country of Prester John, and attention turned again to Central Asia.

Long before the Portuguese arrived in India, a pre-existing Christian presence in the East was a known reality. As well as the Churches of the Eastern Rites which existed throughout the Levant and Middle East, the Nestorians had spread further east in Asia and in addition it had been known that the Church of the Apostle Thomas existed in south India. For European scholars, it was not inconceivable that another significant Christian community flourished in the lands of Prester John. Early reports of a non-Muslim emperor, who was defeating Arab armies and conquering large parts of the East, helped rekindle the belief, but this hope was crushed when Europeans were faced with the reality of the Mongol invasions. Despite the absence of any proof of its existence, when the Portuguese set up colonies in the East, they believed they might encounter the fabled land. The expansion of their empire was commercially driven with a concomitant desire to introduce Catholicism to their subject peoples, but missionaries and traders were willing to listen to accounts of lands where the inhabitants followed unknown religions. As the Jesuits spread eastwards to Japan and China, they encountered many such lands, but no Prester John. It was, however, one of the most influential of these Jesuits who instigated the first European expedition to explore Central Asia, motivated in part by the desire to find the mythical Christian nation.

The Jesuit, Matteo Ricci, had been invited to join the court of the Chinese Wanli emperor, Zhu Yijun, in 1601 because of his skills as an astronomer, mathematician and cartographer. Ricci had spent nearly twenty-five years in China working as a missionary before receiving this great honour. His knowledge of the country and its civilisation led him to believe that the fabled Cathay at the end of the Silk Road was China. He wrote to his superior in Goa in 1598 with the suggestion that an expedition be sent from India along the Silk Road with the intention of meeting him in China. As part of their mission, the Jesuit explorers were to seek out Christian communities along the way. A commercial element was added to the venture when Ricci suggested that the overland route might prove easier than the long sea voyage to China. In 1602 a Jesuit lay brother, Bento de Goes, with a small group of companions, left Agra and travelled through Afghanistan and Central Asia. After a journey of more than three years, he and a single companion arrived at Su-cheu, a town guarding a gateway through the Great Wall. By that point in his journey, he was gravely ill and could go no further. He sent a letter to Matteo Ricci in Beijing who sent help, but Goes did not recover and died in Su-cheu in 1607. Using the information that Goes had gained on his travels, Ricci wrote a report describing the parts of Central Asia he had explored, including the peoples he had met and their customs. He reported that there were communities of Chinese Buddhists, whose existence he believed had given rise to the rumours of peoples who followed Christian practices, but Goes had found no Christian communities (Wessels 1924: 1–41). The expedition had succeeded, to the extent that it had established beyond doubt that Cathay was China, but the land of

Prester John had not been found, and the overland route through Central Asia to China presented a much more difficult trading route than the alternative sea voyage. The Jesuits made no further attempt to find the land of Prester John for another seventeen years.

Lamas from the West

In 1624 a party of Jesuits set out from Agra to travel to Tibet. They were motivated by accounts that, in the land beyond the Himalayas, there lived a people who followed Christian rituals. Again, those making the reports were confusing the practices of Buddhism with those of Christianity. The Jesuits joined a group of Hindu worshippers who were undertaking an annual pilgrimage to the source of the river Ganges, and travelled with them to their holy shrine at Badrinath in what is now the Indian state of Uttarakhand. It was under the control of the rajah of Srinagar in Garhwal (not to be confused with the much larger city of Srinagar in Kashmir). The rajah tried to prevent the Jesuits from leaving his kingdom, but the leader of the party, Fr Antonio de Andrade, accompanied by Brother Manoel Marques, managed to cross the Mana Pass, almost 18,500 ft above sea level, and arrived at the Tibetan town of Tsaparang, capital of the kingdom of Guge. They were the first Europeans to cross the Himalayas.

The king of Guge welcomed them and invited them to stay, but they left before winter came and closed the mountain passes for the year. Andrade returned the following year with two companions and with the support of the king, Tri Tashi Drakpa, they built a church and began to evangelise (Allen 2000: 243–5). Over the course of four years, five other Jesuits joined them and expanded their mission presence from Tsaparang to several other settlements in the kingdom. Andrade sent reports back to Goa explaining the political situation and the progress of the missionaries in making conversions. He said little about Tsaparang itself, although later Jesuit reports gave fuller accounts. The town now lies in ruins but was an important trading centre on the Silk Road. It was situated at the head of the river Sutlej, and merchants from Kashmir arrived there once a year and met up with Chinese traders to hold a great market. The Chinese brought raw silk, porcelain and tea which they exchanged for Indian cottons, finished silks and dried fruits (Wessels 1924: 65).

Once the merchants had departed, the town reverted to being the centre of the surrounding agricultural community of livestock herders. The two dominant buildings in Tsaparang were the king's palace and the Buddhist monastery, whose abbot was the king's brother. Despite the family relationship, the abbot resented the king's support for 'the Lamas from the West' and, when Andrade left to return to Goa in 1629, the abbot led a revolt against the king. Tri Tashi Drakpa was deposed with the help of the neighbouring king of Ladakh. Guge was incorporated into Ladakh and the Jesuit mission stations were ransacked. The new king closed the mission churches and held the remaining five Jesuits captive in their house in Tsaparang. He hoped that he could use them to extort a ransom of Western goods, especially firearms. In 1632 another Jesuit, Fr Francis de Azevedo, travelled to Tibet and returned to Agra with news of the situation that the missionaries were enduring and the king's demands. In February 1633, Andrade wrote to the Jesuit General in Rome, Mutio Vitelleschi, requesting permission to form another Jesuit expedition to rescue the trapped missionaries. Andrade hoped that, through diplomatic means, he could get the king of Ladakh to allow him to rebuild the mission in Tsaparang. Vitelleschi approved, but before Andrade could finalise his plans he was murdered in Goa in 1634 by a fellow Portuguese. It fell to Fr Nuño Coresma to lead the expedition of seven to Tsaparang.

It was at this point that George Strachan appears to have become involved with the mission to Tsaparang. In 1634 he was in Iran working on the book on logic, *Jām Gētī Numā.* He appears to have spent the previous ten years earning a little income which enabled him to indulge his academic interests and fulfil his ambition of travelling to Agra. Prospects there were not as he had hoped, and the unfavourable conditions created by the Inquisition in Goa made it prudent to return to Isfahan. There he could enjoy the company of the Carmelite friars and possibly join in discussions with Mīr Dāmād's circle of scholars. This comfortable existence changed following the death of Mīr Dāmād in 1631–2. Strachan's connections with the Persian court must have lessened, if not ended. With nothing to detain him in Isfahan, news that Andrade was preparing an expedition to travel to Tibet would have excited the wanderlust in the Scot.

Strachan was capable of making impulsive decisions: the most life-changing was the one he took in Aix-en-Provence in 1613, to abandon his position in the household of the Duc de Guise and travel east. The exploits of Andrade and his companions had been publicised almost as soon as his letters from Tibet began arriving in Goa. *Relaçam da Missam de Tibet,* although only a brief account, was published in Goa in February 1626. Two years later a fuller account using Andrade's letters was published in Rome in an Italian translation (Franco 1717: 400–3). Due to the great interest that was shown in Europe regarding events in the East, these accounts were widely circulated throughout the Society of Jesus and beyond. On his visits to Goa, Strachan would have been given the latest news from Tibet, but the Carmelites in Isfahan also would have been aware of these events. It appears that these accounts caused Strachan to travel to Goa and accompany Coresma's party when it left Goa in early summer of 1635.

Final Journey

The circumstances surrounding Andrade's expedition of 1624 and that led by Coresma could hardly have been more different. Moghul sentiment towards the Jesuits had turned hostile. In 1632 Shah Jahan's army destroyed the fort and trading post at Port Hoogly in the Ganges delta and drove the Portuguese traders and Jesuit missionaries from Bengal. Hundreds of Christian prisoners were taken in chains to Agra, where the priests were put in prison, a number of them dying there. Other prisoners were forced to convert to Islam or face enslavement (Maclagan 1932: 103–4). Notwithstanding the political hostility, the greatest difficulty Coresma's expedition had to overcome was the widespread famine that had gripped the north-east and Deccan region of India. The monsoon rains had failed in each of the three years from 1630 to 1632 resulting in widespread famine. The problems this caused were aggravated by Moghul demands to feed the army engaged in fighting the Portuguese. Dutch merchants in Surat estimated that nearly seven and a half million people died of starvation (Winters 2017: 134).

When Coresma and his small group set out, the land was still recovering from what has been described as the greatest famine to strike India (Ó Gráda 2007: 5–38). Any food available could only be obtained at great expense and, in the course of their journey, the expedition spent more than 3,000 rupees on provisions (Wessels 1924: 83). By the time they reached Srinagar in Garhwal, two of the party had died and three more were too ill to go any further. Coresma travelled on to Tsaparang with one companion, and quickly came to the conclusion that it was futile to attempt to maintain the mission there. The surviving Jesuits of Andrade's mission along with the small number of Christian converts had scattered. Coresma and his companion were held captive before being released to return to India. Several further attempts were made to restart the mission in Tsaparang but were brought to an abrupt end in 1640 with the capture of Brother Manuel Marques by the Tibetans. The Jesuits attempted to obtain his release and a missionary presence was maintained in Srinagar to support these efforts. They obtained the support of the father of the Rani of Lahore, who sent a letter to the Tibetan king recommending the Jesuit's release. The last that was heard of Marques was when he sent a letter in 1641 pleading for help. The Jesuits were not allowed to venture into Tibet, and the last missionary withdrew from Srinagar when all hope had gone of Marques still being alive. No reference to the mission is to be found after 1656 but its closure could have been as early as 1644 (Wessels 1924: 87–8).

Although the mission at Srinagar was short-lived, one artefact survives which testifies to its existence and George Strachan's involvement. The printed copy of the New Testament in Arabic that Major William Yule acquired in India at the beginning of the nineteenth century (Chapter 6) contains two inscriptions. The first, written by Strachan, explains that he had read it in its entirety in only twenty days in 1616 while travelling in the desert as part of Emir Feyyād's household. The second inscription on the flyleaf is shorter: *Missionis Xrinagarensis* (belonging to the Srinagar Mission) and is written in a different hand (Maclagan 1932: 354). The mission was never intended to be permanent and would not have warranted the establishment of a library. Any books it possessed would have been gained accidentally. There is no record of how Strachan's New Testament came to be in Srinagar. It must have arrived at some point between 1635 and 1644, the period when the Jesuits were based in the city rather than simply visiting it. There are a number of possible ways in which it might have been brought there. It may have come into the hands of a Portuguese Jesuit who could read Arabic, but there would have been no reason for him to take it on an expedition to a mission station in Tibet through lands where no Arabic speaker lived. The most plausible explanation is that from 1616 it remained in Strachan's possession as his vade mecum, and that it was still in his possession in 1635 when he joined Coresma's expedition.

Since the book was left in Srinagar, it would indicate that Strachan was one of the two members of the party who did not survive the arduous journey and died before they reached Srinagar. Jesuit historians have researched carefully the accounts of the mission to Tsaparang. Through cross-references with Jesuit archives in Rome, they have been able to identify all the missionaries who participated with the exception of the two who died (Desideri 1728). Identification of the members of the expedition was possible due to the detailed archives in Rome. These archives suffered following the dissolution of the Society in the later eighteenth century but when Ippolito Desideri wrote his history the records were intact. His failure to identify the two fatalities on Coresma's expedition is most likely to have been due to their not being members of the Society of Jesus. Having non-Jesuit members of the party would not have been exceptional. Bento de Goes had started his epic journey to China with a mixed group of Jesuits and others. His sole companion on reaching the Great Wall was an Armenian merchant who had traded on the Silk Road and was valued for his experience. It is likely that George Strachan joined Coresma's expedition in the capacity of a physician, and the Roman archives would have had no record of him. His death on the journey to Srinagar was recorded in the reports of the expedition but not his or the other fatality's name. Despite their own physical distress, the Jesuits who accompanied Strachan would have given him a Christian burial. His belongings may have been abandoned but the copy of the New Testament in Arabic would have been kept out of respect for the holy text. It appears that they took what they could but on reaching Srinagar they left the book with the three members of the group who were too ill to journey any further to Tibet.

George Strachan was sixty-three years old when he joined Coresma's expedition in Goa. Having spent his life travelling, he was inured to its hardships and was physically fit. Nevertheless, his companions would have considered him to be elderly and, although the concept would have been unknown to him, his body still harboured the malaria parasite. The journey to Tibet proved to be particularly arduous and, along with the others, he struggled with conditions of near starvation. His companions were at least twenty years his junior (Wessels 1924: 83–7) but even they were exhausted by the time they reached Srinagar. It is not surprising that Strachan should become a casualty in such harsh conditions: his advanced age and susceptibility to recurrences of malaria left him vulnerable. The sweltering pre-Monsoon summer heat of the Indian plains would have reinforced his view that the disease was caused by 'bad air'. Weakened by starvation, Strachan is likely to have been subjected to a debilitating attack. He would have had little resistance and, despite his supply of medicines, it is likely that he succumbed to a recurrence of the disease. As an exile from his native land, it was inevitable that George Strachan would not end his days in the Mearns, but he could not have imagined his final resting place being so different from Scotland. However, given his lifelong interests, it was fitting that his death and burial on the plains of north-west India was in the company of Jesuits.

Strachan's Legacy

The place and time of Strachan's death may help explain the most serious problem faced by researchers: the almost complete lack of any records of his life in the East provided by Strachan himself. The copy of his will held by the Carmelite order in Rome and his letter to Sir Thomas Smyth of the East India Company in London are written in his hand. Together with the annotations he made in his book collection, these are all that survive of Strachan's writings during his travels in the East. While in Aix-en-Provence in 1613, it was his discussion with William Lithgow that had inspired his decision to go east. Lithgow was funding his expeditions by publishing accounts of the journeys, and Strachan must have considered doing the same. When he met Pietro Della Valle in Persia, his friend informed him during their many conversations that he was writing at great length to Dr Schiapano in Italy with the intention of publishing an account of his travels. This would have reinforced Strachan's belief that any book describing his own adventures would have been equally interesting and potentially profitable. However, he could not have produced such a volume without keeping a diary or making comprehensive notes. No such records have survived.

When he entrusted his library to Fr Vincent on his return to Rome in 1622, no mention was made of diaries or notes. Strachan must have retained these, perhaps with the intention of working on a travelogue. When he journeyed to Goa in 1635, it is likely that he left manuscripts with the Carmelites in Isfahan. The Vatican library has a number of Eastern manuscripts which are marked as originally belonging to the priory in Isfahan. Professor Dellavida has suggested that some of these may have belonged to Strachan (Dellavida 1956: 100–1). On arrival in Goa, he may have decided to leave all non-essential items in the safekeeping of St Paul's College. If so, they would have been destroyed in the catastrophic fire that engulfed the college in 1664. The contemporary Jesuit historian of India Fernao de Queiro recorded that the fire consumed almost the whole of

St Paul's College library and its collection of archives (Županov 2019). Strachan's travel notes may have been deposited in the library of St Paul's College, left at his graveside on the Indian plain or taken to Srinagar. In each case they would not have survived.

The loss of these notes and any narrative that Strachan may have constructed from them has deprived researchers of an invaluable resource. The extent to which he appreciated the culture of the societies that he studied would have been manifest rather than being left to reasoned argument. It is clear that he did not return to Europe, as some early researchers believed possible; however, doubt remains concerning many other aspects of his life. Did Edward Monnox involve Strachan in his negotiations with Shah Abbas and Quli Khan regarding English participation in the attack on Hormuz? Was Strachan a member of Mīr Dāmād's circle of scholars? Did he take part in the Jesuit expedition to Tsaparang in 1635 and was he one of the fatalities on the journey? Interesting as it would be to have certainty on these points, in a real sense it is not important. His legacy rests on the fact that he dispatched his collection of books to Rome in 1622. Strachan's library formed a repository that was used by later European scholars and contained books which at that time were unknown in the West. Those texts which previously had been available in Europe were usually imperfectly understood. Strachan's insightful translations and commentaries advanced Western understanding of Eastern culture and religion in ways that had been impossible before. The work of Ludovico Marracci and those who followed him would have been greatly impoverished, if not made impossible, without Strachan's interpretation of texts. His work on *I'lām al-wará bi-a'lām al-hudá* ('Teaching human beings the signs of the right path'), the writings on the life of Mohammed by the Shi'a scholar Raḍiyyaddīn al-Faḍl ibn al-Ḥasan at-Ṭabarsī, was groundbreaking in Europe. Some of Marracci's most important publications can be considered to be ones that Strachan would have produced had he returned to Europe.

The poverty of understanding of seventeenth-century European scholars prior to Strachan can be ascribed to the manner in which they gained their knowledge of Middle Eastern culture and Islamic matters. There were few native speakers of Arabic that scholars could consult as teachers. Even Etienne Hubert, who taught classes in the University of Paris for a few years and had lived in North Africa for a short time, was exceptional (Lefranc 1893: 383). The proficiency in the language that such teachers acquired did not extend to an adequate understanding of its literature and culture. Similarly, the religious orders in Rome, most notably the Order of the Caracciolini, who taught Eastern languages, restricted their intake of students to those intending to work on the Christian missions. Rather than study Eastern culture, which they believed to be flawed due to its dependence on the 'false' religion of Islam, their efforts were concentrated on refuting if not denigrating all Islamic ideas (Zwartjes 2012: 185–242).

With the restrictions on the quantity and the questionable quality of the tutors available, the few students of Arabic who existed in Europe were forced to rely heavily on Arabic texts to pursue their studies. But these were scarce and poorly understood. Erpenius' successor at Leiden, Jacobus Golius, could only publish the first printed copy of poetry in Arabic script as late as 1629 (Loop 2017: 233). Students were better served by the works of a number of Arabic scholars who had converted to Christianity. In the sixteenth century Juan Gabriel and Leo Africanus were prominent among these who, although they did not teach, published a few works which were valued by scholars (Martin 1992: 157). Cardinal Viterbo commissioned Gabriel and Africanus to produce a Latin translation of the Qur'an but, for doctrinal reasons, he would not publish it. One of Africanus' published works, *Description of Africa* (1600), included a series of poetic funeral orations which were quotations from classical Arabic texts. Erpenius valued these as examples of Arab poetry which European scholars could study knowing that their understanding was correct in both meaning and context (Loop 2017: 233). But they had few other texts that they could read with the same confidence.

When viewing European understanding of Arab literature and culture in this light, it is easier to understand the importance of Strachan's library. In themselves, the books are valuable but their real treasure lies in the wealth of erudition that Strachan added in his translations and glosses. He had gained his insight through his exposure to the intellectual culture that produced the works. Strachan had spent more than two years in the household of Emir Feyyād, at the same time as engaging in intellectual argument with his Ḥusaynābādī teachers. He followed this period of intensive exposure to Arabic culture by exploring the libraries of Baghdad, studying a large number of Arab, Persian and Turkish texts before deciding which purchases he wanted to add to his eclectic selection. Through his total immersion in the language and culture of his host community, he gained an understanding that was impossible to attain in the classrooms of Europe.

In the process, Strachan himself changed. When he presented himself to the East India Company merchants in Isfahan in 1619, Thomas Barker described him in the words 'hee is become as well a French man as a Scotish man and verrie little or nothing at all an English man' (Yule 1888: 322). Pietro Della Valle also saw the refined Scottish gentleman, but perhaps came closer to understanding the change that was happening in Strachan when he wrote 'the very Arabs do not distinguish him from a genuine Bedouin' (Della Valle 1664: 580). His life among the Arabs in Syria and Iraq and in Shah Abbas' Iran was more congenial to him, culturally as well as intellectually, than the one he had lived as an impoverished scholar in Europe. Philosophical discussions with Mīr Dāmād and his circle of friends, either directly or through their writings, appear to have enriched still further his life in the East. After fleeing Emir Feyyād, he had been unable to retrace his steps westward but, in the course of the following seventeen years, he had opportunities to return to Europe and did not take them. It would appear that he had fallen in love with the East. His initial intention to return home was forgotten or indefinitely postponed. His change of plan was made in the knowledge that his library in Rome would serve the purpose that he had stipulated in his will:

> I spontaneously and freely give and piously bequeath as alms all the aforesaid books to the Monastery of the Saint Paul Mission of the Discalced Carmelites in Rome, so that the Fathers and Monks of the monastery may pray for my soul and the atonement of my sins in their masses, orisons and penitences.

Having taken care of his soul, George Strachan was free to remain in the East following his life as a scholar and traveller in exotic lands.

Appendix

The following is a list of known manuscripts belonging to Strachan's library, using Strachan's catalogue numbers, where appropriate. Dates are given in both the Hijri and Gregorian calendars. The diacritics shown for the transliteration of Arabic and Persian script are those used in the given quotations. Dellavida uses DMG (Deutsche Morgenländische Gesellschaft). Others use ALA-LC (American Library Association and Library of Congress).

Strachan's Catalogue No.: 5 Vatican Library Arabic: 422
Title: *Bānat Su'ād*
Author: Ka'b ibn Zuhayr
Strachan's comments: Poema magni nominis inter Arabas, cui nomen [Burda], Auth(ore) Kiab ibn Zuheir qui tempore Mahometis floruit et hoc poema et ipse pseudopropheta discendu(m) suis commendare solebat et lamiat il arab appellabat quia in hanc literam o(mn)es versus desinu(n)t. In singular carmina commentaries [opera?] Ibn Hisham [grammati]ci eximij. Emit Babilonj An(n)o Dni. 1619 Georgius Strachanus Merniensis Scotus.

A poem greatly appraised among the Arabs, entitled *Burda*, the author of which is Kiab ibn Zuheir, who flourished in Mahomet's time. The Pseudoprophet himself recommended that his followers learn this poem and called it *lamiat il arab*, because all its verses end with that letter. Each verse has a commentary by Ibn Hisham, an outstanding grammarian. Bought at Babylon in AD 1619 by George Strachan, of the Mearns, Scot.

Dellavida's comments: The story related with Ka'b ibn Zuhayr's poem is famous in the history of early Arab literature. When Mahommed had founded his community in Medina but not yet achieved his triumph by the conquest of Mecca, Ka'b ibn Zuhayr, who at that time was one of the best known poets of Arabia, attacked him and his new religion in a violent satire. The increasing power of Mahommed made the position of the poet very dangerous. He was threatened with death should he fall into the hands of the offended Prophet. Therefore, he resolved to appease him, and went directly to the lion's den, to present another poem to him, in which he apologised for the previous attacks, and extolled Mahommed and Islam. Not only was he forgiven but he had the honour of being given the Prophet's own mantle, which was later sold to the Caliphs by his heirs and which became one of the most holy relics in the Islamic world. The poem itself received its title from the mantle (in Arabic *al burda*), the same title was also given to a no less famous ode by a much later poet, al-Būsīrī (cat. No. 36) but also and perhaps more often, referred to as the *Bānat Su'ād* poem, from its initial words '*Su'ād* (the beloved) has departed . . .'

Numerous commentaries were composed on the *Burda*, the most famous of which is by Jamāladdīn 'Abdallāh Ibn Hishām, an Egyptian philologist who died 761/1360. It was first published from this very manuscript (Strachan's) by Ignazio Guidi (Rome, 1871–4) who quotes the Assemani-Mai Catalogue in which Strachan's endorsement is copied but any mention of his name is omitted.

Strachan has been fairly correct in his statement on that famous poem. However, the title *Lāmīyyat al-'Arab* was never given to Ka'b's *Bānat Su'ād*, although it is actually an *l*-poem, but to another famous poem by the pre-islamic poet ash-Shanfará; Strachan must have been misled by some of his Arabic teachers.

The manuscript is dated 18 Dhu'l-ḥijja, 988/24 January 1581. It was less than forty years old when Strachan purchased it.

Strachan's Catalogue No.: 7 Vatican Library Arabic: 394
Title: *Kitāb al-anwār li-a'māl al abrār* ('The Book of Lights for the acts of religious people')
Author: Yusaf al-Ardabīlī
Strachan's comments: Hic est liber Luminum, id est Universum Corpus Juris Sectae Mahometanae secundum ritum Shafeiorum, Liber amplissimus et curiosissimus.

This is the book of the Lights, i.e. the whole Corpus Juris of the Mahometan sect according to the Shafeit rite, a most wide and curious book.

Dellavida's comments: A widespread treatise on positive law by Yusaf al-Ardabīlī (d. 776/1364 or 799/1396). Strachan's short description is correct: the Shafi'ite rite (or rather school of religious and civil law) is one of the four recognised as orthodox in Islam.

The manuscript is dated 881/1476–7. At the end, near the colophon, we read a half-blurred eulogy of 'Sultan Ḥasan Beg' who certainly must be identified with the king of Persia, Uzūn Ḥasan, who died in January 1478.

Strachan's Catalogue No.: 8 Vatican Library Arabic: 408
Title: *Jāwīdhān Khiradh* otherwise called *Ādāb al-'Arab wa'l-Furs* ('The Culture of the Arabs and the Persians')
Author: Ibn Miskawayh (d. 421/1030)
Strachan's comments: Strachan made no comments on this manuscript. It is identified as his by the catalogue no. entry being in his handwriting.
Dellavida's comments: A fine calligraphic manuscript dated 11 Muḥarram, 741/7 July 1340 . . . a work of invaluable importance.

Strachan's Catalogue No.: 11 Vatican Library Arabic: 516
Title: *Nahj al-bālāgha* ('The Way of Eloquence')
Author: Alī ibn Abī Ṭāleb
Strachan's comments: Nahaga il balága. Iter Elequentiae, liber in quo continentur, Orationes epistolae, sermons paraenetici et declamations, pace belloque dictae an Aly ibn Aby taleb, patruele et genero Mahumetis, eleganti stilo arabico. Opus apud Arabe inventu rarum. Describi curavit ex antiquiss(i)mo exemplari Babiloni anno Dni. 1618 Georgius Strachanus Merniensis Scotus.

Nahaga il balága. The Way of Eloquence, a book which contains the speeches, letters, exhortative sermons and harangues delivered in elegant Arabic style by Aly ibn Aby Taleb, Mahomet's cousin and son-in-law, in times of peace and war; a work scarce among the Arabs. Copied in Babylon in AD 1618 from a very ancient manuscript for George Strachan of the Mearns Scot.

Dellavida's comments: The description is correct. The *Nahj al-bālāgha* is an ample collection of speeches and letters allegedly pronounced and written by Alī ibn Abī Ṭāleb, the fourth caliph and the first among the Imams, or sacred leaders, venerated by the Shi'a sect. Its authorship is discussed, and the brothers ash-Sharīf al-Murtaḍá (d. 436/1044) and ash-Sharīf ar-Raḍī (d. 406/1026), prominent members of the 'Alī family, are both credited with its collection and edition. This work is widespread both among Sunnites and Shī'ites; therefore Strachan's assertion that manuscripts of it are rare is an unjustified claim. The manuscript obviously written in the year of its purchase bears no date.

It is in its original oriental binding of stamped red leather.

Strachan's Catalogue No.: 13 National Library Naples: 39
Title: *Dīwān*
Author: Wrongly credited to Caliph Alī
Strachan's comments: Diuan, id est palatium, Aly I. Aby Taleb. Est opus diversorum poematum piorum. pietate Araba, inscriptum predicto Aly genero et patruelj Mahometis. Emit Babilonj anno Dnj 1617 Georgius Strachanus Merniensis Scotus.

Diuan, that is palace, of Aly son of Aby Taleb. It is a work consisting of diverse pious poems, according to Arab piety, attributed to the abovesaid Aly, son-in-law and cousin of Mahomet.

Dellavida's comments: Strachan understood the word dīwān as meaning 'palace' (it properly should be 'reception hall'), whereas it obviously means 'collected poems'. Actually, the manuscript contains a Shī'ite collection of poems, entitled *Anwār al-'uqūl li-waṣī ar-Rasūl*, with which the caliph 'Alī is wrongly credited. The manuscript was written at the beginning of Dhu'l-qa'da 1023/December 1614.

Strachan's Catalogue No.: 15 Vatican Library Arabic: 461
Title: *Ḥalbat al-kumyat*
Author: An-Nawājī
Strachan's comments: Helbet ilkumeit. Emulsio vini, Liber rarus, multis historiis, fabulis facetis, diversorumque Arabum poematis (sic) constans, in laudem Vinj, et vitae lautioris, convivandique optima ratione. Emit Babiloni a.D. 1618 Georgius Strachanus Merniensis Scotus.

Helbet ilkumeit. The mixing of wine, a rare book, consisting of numerous stories, witty tales and poems by various Arabs, in praise of wine and luxury, and dealing with the best way of banqueting.
Dellavida's comments: *Ḥalbat al-kumyat* by An-Nawājī (d. 859/1455), a book very popular in Arab literature, although its subject made it unlawful. Strachan failed to understand the pun in the title: Ḥalbat al-kumyat means both 'The race track of the bay horse' and 'The milking (i.e. pouring) of red wine'. As for the rest he was quite correct.

The manuscript was apparently written in Egypt. It is dated 18 Muḥarram, 864/14 November 1459, and is written in calligraphic handwriting. Its title page is nicely illuminated. However, we possess other manuscripts of the *Ḥalba* which were written during the author's lifetime, and even by his own hand.

Strachan's Catalogue No.: 16 Vatican Library Arabic: 583
Untitled miscellaneous poetry
Strachan made no comments on this manuscript but the catalogue no. entry is in his handwriting.
Dellavida's comments: A miscellaneous manuscript, mostly of poetry, and written on Persian paper.

Strachan's Catalogue No.: 18 Vatican Library Arabic: 510
Miscellaneous works including *Shihāb al-akhbār, Maqṣūra, at-Tarjī'iyya, Badā'I' az-zamān fī waqā'I' Kirmān.*
Authors: Mohammed, by al-Quḍā'ī (Abū 'Abdallāh Muḥammad ibn Salama), Ibn Durayd, al-Mūsawī, aṭ-Ṭanṭarānī, Abū Muḥammed ibn Mukhtār al-Asīl and Afḍaladdīn Ibn Ḥāmid.
Strachan's comments: Hoc libro haec continentur. Primo opus mille sententiarum, piae quidem, illae aut morales. Distinctum in praefationem, 17. Capita et Conclusionem Authore Il Cadi Abu Abdalla Mahomete Egyptio. 2. Poema compositum a principe quodam Babilonio, Amid Il daula cum esset Detentusin Arce Iskanderana a Chalifo Babiloniae, habet suas notas marginales et interlineares quae usui Comentarij esse possunt. 3. Poema quoddam celeberrimi poetae et antiqui nomine Ibn Il Warid Cuius Initium Ia Dhabiata. Cum notis marginalibus et in fine libri eleganti comentario. 4. Opus variorum poematum pulcherrimorum, quibus titulus est Il Hhagiaziat (sic) hoc est Carmina provinciae Hagiazae, quae Meccam et Medinam capit authore Il Cadi il Moŭsawj. Septem capitibus distinctum. 5. Poema breve

sed celebre inter arabas doctos ob frequentes vocum in eo paronomasias in singulis suis versibus id est duplex vocumin sensu et sono allusion cum comentario ad singula Carmina . . . eius initium Ia chali il balj. Item in fine eius habetur Apendix Historiae Chaliforum Babiloniae Lingua Persica. 9. Capitibus innominato authore.

In this book the following works are contained: First: A work of one thousand sayings, either pious or moral, consisting of a preface, seventeen chapters and a conclusion, written by the Cadi Abu Abdulla Mahomet the Egyptian. 2. A poem composed by a certain Babylonian prince, Amid il daula, when he was held up by the Caliph of Babylon in the fortress of Alexandretta. It has marginal and interlineal notes, which may be used as a commentary. 3. A certain poem by the most famous and ancient poet Ibn Il Warid, beginning with the words 'Ia Dhabiata' with marginal notes and an elegant commentary at the end of the book. 4. A work consisting of various most beautiful poems, entitled Il Hagiaziat, i.e. the poems of the province Hagiaz, which comprises Mecca and Medina, by the Cadi il Mousawj, divided into seven chapters. 5. A short poem, famous among the learned Arabs on account of the frequent paronamasias occurring in each of its verses, i.e. a double coincidence of meaning and sound in the words, with a commentary on each verse [blurred word] It begins 'Ia chali il bali'. At the end there is the Appendix to the History of the Caliphs of Baghdad, in Persian, in nine chapters, by an unnamed author.

Dellavida's comments: Strachan neglected to enumerate, or failed to recognise, a few shorter items in this miscellaneous manuscript of Persian provenance. The first listed item (actually the first in the manuscript) has the title *Shihāb al-akhbār, 'The Falling Star of Traditions'*, a most reputed collection of sayings attributed to Mohammed, by al-Quḍā'ī (d. 454/1062), his full name is Abū 'Abdallāh Muḥammad ibn Salama. He actually was a judge ('qāḍī) in Egypt. In his original plan, his sayings should have been one thousand, as is stated in the preface, on which Strachan is dependent, but he added later two hundred more. The chapters are sixteen and not seventeen. Fol. 39–47 have been neglected by Strachan: they contain nothing essential. Item 2 is represented only by this manuscript, as far as I know. 'Amīdaddawla ibn Jāhir, a vizir to the caliph al-Muqtadī (by Babylon Strachan of course means Baghdad), was incarcerated in 493/1100 and died in prison soon after. 3 is the famous *Maqṣūra* (a poem with ā rhyme) of Ibn Durayd (d. 321/934), a poem more rhetorical than artistic. The initial words 'Yā Ẓabyata' mean 'O gazelle . . .' In the Arabic script, Durayd may easily be misread Varid and Strachan actually read it thus. Evidently the name of Ibn Durayd, an outstanding grammarian and philologist, was unknown to him. However, the name of the author does not appear in our manuscript, and Strachan must have taken it from elsewhere, in the wrong reading. I have no information at all about the anthology listed under number 4 and in my *Elenco* I overlooked the title and the name of the author (al-Mūsawī), which appear at the end. 5 is an anonymous commentary on a poem by aṭ-Ṭanṭarānī (d. c. 480/1087) in praise of the Seljuk emir Niẓāmaddawala; it is known under the title *at-Tarjī'iyya*, 'The Echoing Poem', because the last and penultimate words, or groups or syllables in each verse are identical in sound, although different in meaning. The beginning of the poem 'Yā khaliyya 'l-bālī', 'O thou who art lonely . . .', was fairly well transliterated by Strachan, who, however, omitted the following item (8), a commentary by Abū Muḥammed ibn Mukhtār al-Asīl on Ibn Durayd's *Maqṣūra*. 6 is not an appendix to a chronicle of the Baghdad caliphs but an appendix to the history of the Persian province of Kirmān (*Badā'I' az-zamān fī waqā'I' Kirmān*, 'The marvels of time on the events in Kirmān') by Afḍaladdīn Ibn Ḥāmid.

The manuscript is old (the last item is dated 15 Sha'bān, 763/ 9 June 1362) and undoubtedly of Persian origin, as is shown by the last item and by the presence of numerous Persian notes throughout. However, it was brought to Medina, where it was in the year 1013/1604, according to a note on fol. 1c, and from there it must have passed to Baghdad where Strachan bought it.

Strachan's Catalogue No.: 19 Vatican Library Persian: 43
Title: *Gūlistān*
Strachan's comments: Hic est Guiliestan, seu Hortus Persicus Variorum Discursuum Eleganti sermone persico scriptus: Emit Babilonij anno. Dni. 1617. Georgius Strachanus Merniensis Scotus.

This is the Guiliestan, or the Persian Garden of different speeches written in elegant Persian.
Dellavida's comments: This is the only Persian manuscript of Stachan's in the Vatican Library, a copy of the *Gūlistān*, 'Rose Garden' of Sa'dī, a collection of ethical stories and verses which enjoyed a tremendous popularity. It is late, perhaps contemporary with Strachan, and not particularly good.

Still in its original binding of cream cardboard and red leather.

Strachan's Catalogue No.: 21 Vatican Library Arabic: 419
Title: *al-Mirāḥ fī sharḥ Marāḥ al-arwāḥ* (the ms. is untitled)
Author: Ḥasan Pāshā ibn 'Alā'addīn al-Aswad
Strachan's comments: Hic est commentaries clarus In librum Aetymologiae Arab(icae) Cui Zingiani est nomen, seu Requies animorum. Emit Babilonij anno Dni 1618 Georgius Strachanus Merniensis Scotus.

This is a famous commentary on the book of Arabic Etymology entitled Zingiani, or the relaxation of the spirits.
Dellavida's comments: The manuscript which bears no title contains the *al-Mirāḥ fī sharḥ Marāḥ al-arwāḥ* by Ḥasan Pāshā ibn 'Alā'addīn al-Aswad (d. c. 800/1397), a commentary on the short text book on grammar *Marāḥ al-arwāḥ*, 'The relaxation of the spirits' by Ibn Mas'ūd (d. beginning of 8th/14th century). Obviously Strachan was misinformed: az-Zanjānī (d.

655/1257) is the author of a famous text book on morphology, *at-Taṣrīf al-'izzī.*

The manuscript is dated 857/1453.

Strachan's Catalogue No.: 23 National Library Naples: 35
Title: *Shir'at al-Islām*
Author: Muḥammad ibn Abī Bakir Imāmzādeh ash-Sharġī
Strachan's comments: Hic liber continet Ritus et Consuetudines populorum, qui sectam Mahometanum sequuntur in 48. Capita digestus a Mohamete fil. Abubekir il Mofti. Emit Babilonij anno Dnj 1617 Georgius Strachanus Merniensis Scotus.

This book contains the rites and customs of the people who follow the Mahometan sect: it was arranged into 48 chapters by Mahomet son of Abubekir, the Mofti.
Dellavida's comments: Strachan's comments are fairly correct. The manuscript contains a treatise on religious duties, *Shir'at al-Islām,* by Muḥammad ibn Abī Bakir Imāmzādeh ash-Sharġī, Mufti in Bukhārā (d. 573/1177).

Strachan's Catalogue No.: 24 National Library Naples: 21
Title: *Tafsīr gharīb al-Qur'ān*
Author: al-'Uzayrī
Strachan's comments: Haec est Explicatio dilucida omnium vocum antiquarum et contextuum obscuriorum per totum corpus Alcorani Dispersorum Ordine Alphabetico digesta ut facilius in promptu singular reperiantur. Opus summè utilissimum iis qui verborum proprietatis et Arabicae linguae interiora scrutantur Authore Abubekir f. Aziz Segestano. Emit Babilonij anno Dni 1618 Georgius Strachanus Merniensis Scotus.

This is a clear explanation of all ancient words and obscure passages scattered throughout the whole body of the Koran, arranged according to alphabetical order so that each of them may easily be found. A work exceedingly useful to those who examine the meanings and idioms of the Arabic language, by Abubekir son of Aziz, from Segestan.
Dellavida's comments: *Tafsīr gharīb al-Qur'ān* by al-'Uzayrī, a native of the Persian province of Sijistān (d. 330/941): a very well known lexicographical study of the Koran. Strachan's evaluation of the book is correct.

Strachan's Catalogue No.: 25 National Library Naples: 92
Title: *ash-Shamsiyya*
Author: al-Kātibī
Strachan's comments: Haec est Logica Arabum ab homine mirae inter eos subtilitatis conscripta, cui nomen erat Shemsi. Emit Babilonij anno Dnj 1618 Georgius Strachanus Merniensis Scotus.

This is an Arab logic, written by a man of admirable subtlety among the Arabs, whose name was Shemsi.
Dellavida's comments: The famous text book on logic contained in this

manuscript is not by an author called Shemsi but its title is *ash-Shamsiyya*, which means 'The Solar Epistle' (i.e. 'dedicated to Shamseddīn', which was the honorary appellation, meaning 'the Sun of Religion', of Muḥammad ibn Muḥammad al-Juwaynī). The author's name is al-Kātibī, a famous scholar of Persian birth, who died in 675/1276 or 693/1294.

Strachan's Catalogue No.: 26 Vatican Library Arabic: 466
Title: *Taḥrīr al-qawā'id al-manṭiqiyya fī sharḥ ar-risāla ash-Shamsiyya*
Author: Quṭbaddīn at-Taḥtānī
Scribe: Sulṭan Bāyezīd ibn Ḥasan 'Ali
Strachan's comments: Hic est Commentarius clarissimus et Usitatissimus apud Arabes In logicam illam subtilem Viri ingeniosi Il Shemsi Emit Babilonij an(n)o Dnj 1618 Georgius Strachanus Merniensis Scotus.

This is a commentary most famous and much used among Arabs, on the subtle logic of the ingenious man Il Shemsi.
Dellavida's comments: A commentary on the preceding work (Cat. No. 25), entitled *Taḥrīr al-qawā'id al-manṭiqiyya fī sharḥ ar-risāla ash-Shamsiyya*, 'The setting up of the rules of logic in commenting *The Solar Epistle*', and composed by Quṭbaddīn at-Taḥtānī (d. 766/1365).

The manuscript was written on 20 Dhu'l-qa'da 965/ September 3 1558 in Aḥmadnagar, a city in Western India, by a Persian born in or originating from Meshhed, Sulṭan Bāyezīd ibn Ḥasan 'Ali.

Strachan's Catalogue No.: 27 National Library Naples: 20
Title: *Taqrīb an-nashr fi'l-qirā'āt al-'ashr*
Author: Muḥammad ibn Muḥammad al-Jazari
Strachan's comments: Huic libro nomen est Tacrib Jnnishare fi Karaat il ashre. Id est Collectio amplitudinis in decem (Alcorani) lectionibus. Est Compendium Dilucidum De recte Arabum pronunciatione authore Mahomete f. Gizri. Emit Babilonj anno Dnj 1618 Georgius Strachanus Merniensis Scotus.

This book is entitled Tacrib Jnnishare fi karaat il ashre. That is a collection of amplitude concerning the ten readings (of the Koran). It is a clear summary about correct pronunciation, by Mahomet son of Gizri.
Dellavida's comments: *Taqrīb an-nashr fi'l-qirā'āt al-'ashr* by Muḥammad ibn Muḥammad al-Jazari (d. 833/1429), a summary of his major work on the variants of the Koran. Strachan understood the subject of this book fairly correctly, although the word 'qirā'a' refers to variants in writing rather than in pronunciation (see Strachan Cat. No. 37) but failed to grasp the meaning of the first part of the title, which sounds 'Making what is scattered to become near'.

The manuscript is accurately written and considerably old, since it is dated 18 Muḥarram 832/ 29 October 1428 or a year before the author's death.

Strachan's Catalogue No.: 29 Vatican Library Arabic: 447
Poetic works of Ibn Ḥamdīs
Author: Ibn Ḥamdīs whose full name is Abdaljabbār ibn Abī Bekr
Strachan's comments: Hic opera poetica sunt, nobilis et vetutissimus poetae Abd Il gebbar. f. Hemdîs. Emit Babilonj anno Dni. 1618 Georgius Strachanus Merniensis Scotus.

Here are the poetical works of the famous and most ancient poet Abdi l Gebbar, son of Hemdîs.

Dellavida's comments: This is the oldest and most precious of Strachan's manuscripts. As its owner correctly stated, it contains the diwan of the Sicilian poet, Ibn Ḥamdīs, his full name 'Abdaljabbār ibn Abī Bekr' (d. 527/1132), of which only two other mss are known so far.

The manuscript is dated 28 Muḥarram 607/22 July 1210. It was written in Spain in a very clear and correct Maghrebi handwriting, and passed through several hands before it reached Strachan . . . Probably the precious manuscript lay in Baghdad from the 15th to the 17th century, when Strachan bought it.

Strachan's Catalogue No.: 30 Vatican Library Arabic: 404
A commentary on the *al-Kāfiya*
Strachan's comments: Haec est Litteralis seu Gram(m)aticalis Declaratio Libri Syntaxeos Arabicae cui nomen Cafie. Emit Babilonj anno Dnj 1616 Georgius Strachanus Merniensis Scotus.

This is a literal, or grammatical explanation of the book on Arabic Syntax entitled Cafie.

Dellavida's comments: A commentary on the *al-Kāfiya*, the famous text book by the Egyptian scholar, Ibn al-Ḥājib (d. 646/1249). The author is unknown. The manuscript is dated Ramaḍān 1013/ January-February 1605.

Still in its original binding.

Strachan's Catalogue No.: 33 Vatican Library Arabic: 544
A commentary on the *al-'Aqā'id al-' Aḍudiyya*
Author: ad-Dawwāni
Strachan's comments: Discursus subtilis de Veritate sectae Traditionariae, in lege Mahometanae (sic), contra praesularios seu Imâmios (pugna est de lana caprina). Emit Babilonj anno Dnj. 1617 Georgius Strachanus Merniensis Scotus.

A subtle discourse on the veracity of the Traditionalists' sect concerning Mahomet's law, against the priestly or Imâmi sect (a futile contention).

Dellavida's comments: Rather than polemics against the Shī'a, this manuscript contains a commentary on the *al-'Aqā'id al-'Aḍudiyya*, a short creed by al-Ījī (d. 756/1365), the author of which is ad-Dawwāni (d. 907/1501). The terms 'Traditionalists' for the Sunnī and 'Priestly' for the Shī'a are not entirely correct; however, they tally with the general standing of the two great denominations in Islam. 'Imāmī' is quite correct as a designation of the Shī'a.

The manuscript is of Persian origin; it was written in Kāshān in Western Persia on 21 Rabī'II 939/ 20 November 1532.

Strachan's Catalogue No.: 34 Vatican Library Arabic: 580
Title: *as-Sab' al-' Alawiyyāt, Khuṭbat al-bayān, Arba'ūn ḥadīth fī faḍā'il'Alī*
Author: Ibn Abi'l Ḥadīd – see below
Strachan's comments: In hoc libro habentur. P(ri)mo septem poemata in lau(dem) Mahometis Maledicti, et quorundum eius sectatorum, stilo perpulcro conscripta, cu(m) adiunctis com(m)ent(ariis). 2. Monita, seu praecepts, quae fabulantur Shiaej, suum prophetum morientem com(m)isisse Aly ibn Abi talib patrueli et genero suo. 3. Chutbet Ilbeian, sermo Demonstrationis, sunt quaedam assertions fabulosae, et impiae, de Excellentia seu praerogativa Aly ibn Aby Taleb supra totum mundum. 4. Quadraginta discursus fabulosae (sic) de virtutibus et eminentia praedicti Aly Quem Persae totaque secta Shiaej successorem legittimum (sic) Mahometi esse affirmant. Georgius Strachanus.

In this book there are: first, seven short poems in praise of the accursed Mahomet and some of his followers, in a magnificent style, with some commentaries added to them. 2. The admonitions, or precepts, which the Shia pretend their dying Prophet had entrusted to his cousin and son-in-law, Aly ibn Aby Talib. 3. Chutbet Ilbeian. The Sermon of Demonstration, namely fabulous and impious affirmations concerning the excellence, or prerogative, of Aly ibn Aby Taleb over the whole world. 4. Forty fabulous speeches on the virtues and eminence on the abovesaid Aly, whom the Persians and the whole Shia sect affirm to be the legitimate successor of Mahomet.
Dellavida's comments: The Shī'te works enumerated by Strachan are, more correctly, the following: 1. An anonymous commentary on seven poems, known as *as-Sab' al-' Alawiyyāt,* 'The Seven Poems referring to 'Alī', the author of which is the well-known writer Ibn Abi'l Ḥadīd (died at Baghdad 655/1257). Their subjects are some episodes in Mahommed's life, the killing of Alī's son, al-Ḥusayn, and the praise of the caliph al-Nāṣir, who was the author's contemporary. 2. The apocryphal will of Mohammed in favour of Alī, one of the basic texts of the Shī'a sect. 3. *Khuṭbat al-bayān,* another of the basic texts of the Shī'a. 4. *Arba'ūn ḥadīth fī faḍā'il'Alī.* Several variants of these alleged sayings of Mohammed in praise of 'Alī circulated among the Shī'a. Strachan, who seems to have been insufficiently informed about the technical meaning of the word '*ḥadīth*' (an authorised report on something said or done by Mohammed), failed to recognise the contents of the last section in this miscellaneous manuscript, which is very recent, being dated 25 Jumādá II 1027/ 21 May 1618.

Strachan's Catalogue No.: 35 Vatican Library Arabic: 523
A pamphlet attributed to Aristotle
Strachan's comments: Libellus Aristoteli inscript(us) de munere Ducis seu Imperatoris et acibus rite instruendis. Emit Babilonj anno Dnj 1617 Georgius Strachanus Merniensis Scotus.

A pamphlet attributed to Aristotle on the duties of the Chief or Commander and on the proper ranging of the lines of an army for a battle.
Dellavida's comments: Arabic translations of apocryphal letters of Aristotle to Alexander are very common. This one, however, is different from those generally known. The manuscript bears no date: it may belong to the latter part of the 10th/16th century.

Strachan's Catalogue No.: 36 Vatican Library Arabic: 571
A miscellaneous manuscript containing *al-Burda* (formal title *al-Kawākib ad-durriyya fī madḥ khayr al-bariyya*)
Author: al-Būṣīrī
Scribe: Sayyid Ḥājjī (of *al-Burda*; other sections in different hands)
Strachan's comments: In hoc libro haec habentur. P(ri)mo poema nobiliss(imu)m cui no(m)en est Casidet ilburda cum continuo ad illum commentario. 2. Quadraginta fabulae seu discursus morales ad pietate(m) suam fabulosam augendam. 3. Commentarius politus, sed incerti authoris, In poema quoddam Arabicum, cui Lumen nomen dederunt.

In this book are the following items: First, A most famous poem called Casidet ilburda with a continuous commentary. 2. Forty fables or moral talks aimed at the increasing of their fabulous faith. 3. A well polished commentary on a certain Arabic poem called 'The Light' by an unknown author.
Dellavida's comments: The first item in this miscellaneous manuscript is a poem 'qaṣīda' in praise of Mohammed by al-Būṣīrī (d. 694/1294), which enjoyed an unparalleled popularity under the title *al-Burda*, 'The Mantle', with reference to the mantle which Mohammed gave to Ka'b ibn Zuhayr (see Strachan Cat. No. 5). Its full title is *al-Kawākib ad-durriyya fī madḥ khayr al-bariyya*, 'The pearl like stars, in praise of the best of creatures'. In our manuscript it is followed by the commentary by Khālid al-Azharī (d. 905/1499). 2. Another collection of forty Shī'ite hadiths, which Strachan, as usual failed to recognise. 3. An anonymous commentary on the verses quoted as examples of grammatical rules 'shawāhid' in the *ḍaw'*, 'The Daylight' of Tājaddīn al-Isfarā'īnī (d. 684/1285), which is a commentary on a popular text book, *al-Miṣbāḥ fi'n-naḥw*, 'The Lamp on Grammar' by al-Muṭarrizī. Strachan did not catch its character.

This manuscript is the only one purchased by Strachan where his Muslim name is mentioned. A note by the scribe, Sayyid Ḥājjī, states that 'it was copied on behalf of the Frankish physician Mohammed Chelebi'.

Strachan's Catalogue No.: 37 Vatican Library Arabic: 582
Three treatises on the correct reading of the Qur'an and a *ḥadīth*

1. *al-Mufīd fī 'ilm at-tajwīd*
2. *al-Irshād fi'l-qirā'-āt*
3. *Ḥurūf 'Abdallāh ibn 'Āmir al-Yaḥṣubī*
4. a *ḥadīth*

Authors:

1. al-Ḥasan ibn Shujā ibn Muḥammad ibn at-Tūnī
2. Abu'l-Qāsim Manṣūr ibn Muḥammad as-Sindī
3. 'Abdarraḥmān ibn Aḥmad ar-Rāzī
4. Ibn Wad'ān

Strachan's comments: Hic habentur tres tractatus De recta pronuntiatione linguae arabicae. P(ri)mo Il Mufid autore Hassan Tuoano. 2. Liber Reshid De lectione Alcorani. 3. Littera Abu Abdalla f(ilii) Amr De recta Alcorani etiam lectione. 4. In fine habentur 40 fabulae morales. Emit Babiloni anno [Dnj] 1618 Georgius Strachanus Merniensis Scotus.

Here are three treatises on the correct pronunciation of Arabic. First, Il Mufid by Hassan the Tuoan. 2. The Book Reshid on the reading of the Koran. 3. The letter of Abu Abdalla, son of Amr, also on the correct reading of the Koran. 4. At the end 40 moral fables.

Dellavida's comments: 'Pronunciation of Arabic' is not entirely correct (see Strachan Cat. No. 27). Actually, all of the three short treatises contained in this manuscript refer to the correct reading of the Koran, either in the matter of pronunciation 'tajwīd' or of text 'qirā'āt'. The first *al-Mufīd fī 'ilm at-tajwīd*, 'The useful book on the science of "tajwīd"' by al-Ḥasan ibn Shujā ibn Muḥammad ibn at-Tūnī, is extremely scarce since only one other manuscript of it is known so far. Number two, *al-Irshād fi'l-qirā'-āt*, 'The Directory on "qirā'āt"', by Abu'l-Qāsim Manṣūr ibn Muḥammad as-Sindī, is not found elsewhere and the author has not been identified. The third treatise is also known from this unique copy, although its author, 'Abdarraḥmān ibn Aḥmad ar-Rāzī (d. 454/1062), is known. The meaning of the title *Ḥurūf 'Abdallāh ibn 'Āmir al-Yaḥṣubī* was misunderstood and wrongly transliterated by Strachan: Ḥurūf is the plural of 'ḥarf', which actually means a letter of the alphabet, but its technical meaning is 'reading', and refers to the ancient 'readers' who were said to have preserved a text of the Koran different in many instances from the official text. In the fourth item, Strachan, as usual, missed the right meaning of the term '*ḥadīth*'. It is another 'Collection of Forty Traditions' the widespread work of Ibn Wad'ān (d. 494/1101).

The manuscript, not dated, may belong to the 10th/16th century.

Strachan's Catalogue No.: 38 National Library Naples: 67
Three treatises on grammar

1. *Al-Miṣbāh fi'n-naḥw*
2. *Al-I'rāb 'an qawā'id al-I'rāb*
3. commentary on *al-Unmūdhadj*

Authors:

1. al-Muṭarrizī
2. Jamāladdin Ibn Hishām

Strachan's comments: Hic habentur tres Tractatus rari de sintaxi Arabum. P.mus Il Musbach seu Lucerna. 2. Ilierab seu Constructio. 3. Ilenmudhaiḡ, idest opus verbis strictum, sensu largum, De Sintaxi. Cui adjugitur perpetuus Commentarius. Emit Babilonj anno Dnj 1617 Georgius Strachan Merniensis Scotu.

Here are three scarce treatises on Arabic syntax. First, Il Musbach or Lamp. 2. Ilierab or Construction. 3. Ilenmudhaiḡ, or a work narrow in words, and broad in meaning about syntax to which a perpetual commentary is added.

Dellavida's comments: The full titles of those three treatises on grammar, all famous as text books, are: 1. *Al-Miṣbāh fi'n-naḥw* by al-Muṭarrizī (see Strachan Cat. No. 36). 2. *Al-I'rāb 'an qawā'id al-I'rāb* by Jamāladdin Ibn Hishām (see Strachan Cat. No. 5). 3. An anonymous commentary on *al-Unmūdhadj* (this word of Persian origin properly means 'Model'), a compendium by az-Zamakhsharī (d. 538/1143).

Strachan's Catalogue No.: 39 National Library Naples: 68
Four elementary books on grammar

1. *Ash-Shāfiya*
2. *Al-Kāfia*
3. *Al-Ājurrūmiyya*
4. *Al-'Awāmil*

Authors:

1. and 2. Abū 'Amr 'Othmān ibn Omar
3. Ibn al-Ājurrūmi as-Ṣinhājī
4. al-Jurjānī

Strachan's comments: Hic habentur p.mo Liber Shafie de Aetymologia Arabum. 2. Liber Kafie De eorum Syntaxi. Utriusque author Abu Omar Mekanus. 3. Liber Aŭámel: sunt centum regulae breves syntaxeos Arabum.

4. Liber AGerumiae. De Syntaxi Itidem Arabum. Emit Babilonj anno D. 1617 Georgius Strachanus Merniensis Scotus.

Here are, first, the book Shafie on Arab etymology. 2. The book Kafie on the syntax of the Arabs; the author of both is Abu Omaar the Mekkan. 3. The book Aŭámel, consisting of one hundred short rules of Arab syntax. 4. The book AGerumia, also on Arabic syntax.

Dellavida's comments: Four elementary books on grammar. 1. *Ash-Shāfiya* by Abū 'Amr 'Othmān ibn Omar called Ibn al-Ḥājib. 2. *Al-Kāfia* by the same (see Strachan Cat. No. 30). 3. *Al-Ājurrūmiyya,* by Ibn al-Ājurrūmi as-Ṣinhājī (d. 723/1323). 4. *Al-'Awāmil* by al-Jurjānī (d. 471/1078). Strachan's statements are correct, except for the birthplace of Ibn al-Ḥājib, which was Egypt and not Mekka. Abu Omar instead of Abū 'Amr can hardly be called a mistake since the two names spell almost identically in the Arabic script.

Strachan's Catalogue No.: 41 Vatican Library Arabic: 451
Title: *I'lām al-wará bī-a'lām al-hudá*
Author: Raḍiyyadīn al-Faḍl ibn al-Ḥasan at-Ṭabarsī
Strachan's comments: Allām Ilhuda be allām il wara: Signa veritatis per miracula posteriatis, Est historia totius vitae mortis et parentum Mahometis, et filiae eius Fatimae, filioru(m) Aly ibn Aby Taleb ex ea, Nepotum et successoru(m) usque ad decima(m) generationem per 250 et ultra annos, a morte Mahometis, liber iis, qui seriem statumque eoru(m) temporu(m) scire cupiunt expetendus. Transcribi ex antiquiss(i)mo exemplari curavit Babilonj anno Dni 1618 Georgius Strachanus Scotus.

Allām Ilhuda be allām il wara: the signs of truth through the miracles of posterity. It is the whole history of the life and death, and of the parents of Mahomet and of his daughter Fatima, of the sons whom Aly ibn Aby Taleb begot of her, and of their grandsons and successors until the tenth generation, for more than 250 years after Mahomet's death. A book which ought to be obtained by those who want to know the events and conditions of their age. It was transcribed from a very ancient copy at Babylon in 1618 for George Strachan of the Mearns Scot.

Dellavida's comments: The correct spelling of the title is *I'lām al-wará bī-a'lām al-hudá.* Strachan was at a loss in reading and transliterating it, as is shown by the corrections which he made, and he failed to understand its meaning, namely, 'Teaching the human beings the signs of the Right Path'. The work bearing this title is a biography of Mohammed and the twelve Shī'a imams, and concludes with the development of the doctrine of the temporary disappearance of the twelfth imām. It is extremely scarce. Although its author is not mentioned in the manuscript, it was possible to identify him as the renowned Shī'a scholar Raḍiyyadīn al-Faḍl ibn al-Ḥasan at-Ṭabarsī (d. 548/1153 or 552/1158) . . . The manuscript purchased by Strachan was used, while it was still with the Carmelites, by the Italian scholar, Ludovico Marracci, who in 1691 and 1698 published the text and a Latin translation of the Koran, preceded by an extensive and learned exposition and refutation of the theological system of Islam . . . The scribe of our manuscript, Ibrāhīm ibn al-Ḥasan, finished it on 6 Rajab 1027/ 29 June 1618 at Baghdad.

Strachan's Catalogue No.: 42 National Library Naples: 88
A collection of poetry – *dīwān*
Author: Ibn al-Fāriḍ
Strachan's comments: Opera poetica Sheik Omar Ibn Il Fared, qui princeps Amantium nuncupatur. Emit Babiloni anno Domini 1618 Georgius Strachanus Merniensis Scotus.

The poetical works of the Sheik Omar Ibn Il Fared who is called the Prince of Lovers.
Dellavida's comments: The dīwān of the Egyptian mystic poet, Ibn al-Fāriḍ (d. 632/1235), a much read work. The appellation *sulṭān al-'ushshāq* 'Sultan of Lovers' is often given to Ibn al-Fāriḍ with reference to his poetry, in which the language of love is used allegorically to intimate the mystic union with God.

Strachan's Catalogue No.: 43 Vatican Library Arabic: 457
Title: *Aṣ-Ṣaḥīfa al-kāmila*
Author: Attributed to Alī but arranged by his grandson Alī Zayn al-Ābidīn
Strachan's comments: Sachifet il kamila, liber perfectus. Est congeris orationu(m) seu precum Devotaru(m) more Arabu(m) Liber inscrip(tus) Alij ibn Abi Taleb. Emit in Babiloni an(n)o Dnj 1618 Georgius Strachanus Merniensis Scotus.

Sachifet il kamila, the perfect book. It is a conglomeration of orisons or devout prayers according to the custom of the Arabs. A book attributed to Aly ibn Abi Taleb.
Dellavida's comments: *Aṣ-Ṣaḥīfa al-kāmila*, 'The Complete Sheet', a famous collection of Shī'te prayers, which are attributed (although probably incorrectly) to 'Alī. They were gathered and arranged by his grandson, Alī Zayn al-Ābidīn.

The manuscript was written in Ṣafar 1027/January–February 1618 by the same copyist, Ibrāhīm ibn al-Ḥasan, as Cat. No 41. Still in original binding.

Strachan's Catalogue No.: 44 Vatican Library Arabic: 567
Titles:

1. *al-Alfiyya fi farḍ aṣ-ṣalāt al-yawmiyya*
2. *ad-Durra an-nafaliyya fi bayān mā fi'ṣ-ṣalāh an-nāfiliyya*
3. *al-Ja'fariyya*
4. unknown pamphlet
5. unknown pamphlet
6. *al-Bāb al-ḥādī 'ashar*

Authors:

1. Muḥammad ibn Makkī al-Āmidī
2. Muḥammad ibn Makkī al-Āmidī
3. Alī ibn 'Abdal'ālī al-Karakī
4. Unknown
5. Unknown
6. Ibn al-Muṭahhar al-Hillī

Strachan's comments: Hic 6. opuscula habentur. Ex ritu Shiaiorum. 1. Il Elfie i. Mille regulae observandae in oratione praecepti. 2. Il Nefeli i. ordo orationis, ex traditione, praeter eam quae est ex praecepto: utriusque auth(or) Muham(m)ed Il Meki. 3. Il giaferie est ampla, orationis faciendae Declaratio, ex praecedentibus duobus opusculis desumpta. Auth(ore) Sheich Aly. 4. 5 et 6 sunt tres tractatus authoru(m) innominatoru(m) De Aeternitate, et Unitate Dei, mundi creatione, reliquisque Religionis Mahometicae, praecipuis fundamentis. Emit Babilonj anno Dnj 1617 Georgius Strachanus Scotus.

Here are six tracts of the Shī'ite rite. Il Elfie, i.e. one thousand rules to be observed in the obligatory prayer. 2. Il Nefelie, i.e. the order of the optional prayers, according to tradition. The author is Mohammed Il Meki. 3. Il giaferie. It is an ample explanation of the manner of performing the prayer, and is taken from the former two pamphlets. The author is Sheich Aly 4. 5 and 6 are three treatises by three unnamed authors, on the eternity and unity of God, on the creation of the world, and on other basic points of Mahometan religion.

Dellavida's comments: The first is *al-Alfiyya fi farḍ aṣ-ṣalāt al-yawmiyya.* 'The work of the thousand precepts on the duty of the daily prayer', by Muḥammad ibn Makkī al-Āmidī, called ash-Shaykh al-Awwal, 'The First Master' (d. 782/1380). 2. is *ad-Durra an-nafaliyya fi bayān mā fi'ṣ-ṣalāh an-nāfiliyya,* 'The gratuitous pearl on the explanation of the features of the optional prayer' by the same author. 3. is *al-Ja'fariyya,* i.e. 'The work going back to the authority of the imām Ja'far aṣ-Ṣādiq' by Alī ibn 'Abdal'ālī al-Karakī (d. 945/1538) who is possibly the author of the two following pamphlets. 6. is the famous *al-Bāb al-ḥādī 'ashar,* 'The Eleventh Chapter' namely of the *Minhāj aṣ-ṣalāh,* 'The right way of prayer', by Ibn al-Muṭahhar al-Hillī (d. 726/1326), one of the outstanding Shīa authorities. The copying of this manuscript was finished on the first day (1 Muḥarram) of the year 1027/29 December 1617. The scribe was as-Sayyid Hājjī, the same who wrote the first item in Cat. No. 36 and the manuscript must have been bought soon after. The binding is of the original red leather with gold stamping.

Strachan's Catalogue No.: 48 Vatican Library Arabic: 475
Poems of Aḥmad ibn al-Ḥusayn Al-Mutanabbī
Strachan's comments: Opera nobilissimi poetae Mutanebbi, qui propter

stilj gravitatem Ingeniique in poesi miram solertiam, Nomen prophetizantis ab Arabus meruit. Floruit sub califis Babiloniae, circa annum Dni Nostri Jesu Christi 950: Emit Babilonj an(n)o Dnj 1618 Georgius Strachanus Merniensis Scotus.

The works of the most famous poet Mutanebbi, who deserved to be called The Prophesying One by the Arabs because of the seriousness of his style and the marvellous skill of his poetical genius. He flourished under the Caliphs of Babylon about AD 960.

Dellavida's comments: Al-Mutanabbī is the surname of Aḥmad ibn al-Ḥusayn, the most famous among Arab poets of the later period. As he was born in 303/915 and died 354/965, Strachan was correct about his time. Al-Mutanabbī spent most of his life in Syria, Mesapotamia and Egypt, and came to Baghdad only at the end of his life. The reason for his surname (literally 'one who behaves like a prophet') is variously given in the original sources: the version adopted by Strachan has the support of good authorities.

The manuscript, not dated, probably belongs to the 9th/15th century. In 922/1516–17 it was in Khuwārizm, a province of eastern Persia.

Strachan's Catalogue No.: 52 Vatican Library Arabic: 527
Title: *Ma'ānī al-āthār*
Author: Abū Ja'far aṭ-Ṭaḥawī
Strachan's comments: Liber Sententiarum seu Responsionum Selectarum (De Religione Muham(m)edij (sic) et Ceremoniis et ritibus) quae ex ipso Muham(m)edis ore accepta credunt Arabes caeterique illius sectae Orientales. Emit Babiloni Anno Dni 1618 Georgius Strachanus Merniensis Scotus.

A book of chosen sayings or responses about the ceremonies and rites of the Mahommedan religion, in which the Arabs and the other Orientals of the same sect believe, having received them from the very mouth of Mohammed.

Dellavida's comments: Strachan finally learned the true meaning of the *ḥadīth* literature (i.e. the collection of sayings and acts attributed to Mohammed), which he had failed to grasp in his previous endorsements. The work contained in this manuscript bears the title *Ma'ānī al-āthār*, 'The Meanings of Traditions', and deals especially with the problem – a very important one for the setting up of the Islamic religious system – concerning certain precepts in the Koran and Tradition which are believed to be superseded by others: the so-called 'principle of abrogation'. Its author, the ḥanafite scholar Abū Ja'far aṭ-Ṭaḥawī, died in 321/933.

This thick volume of 508 leaves bears no date [but must be earlier than 652/1254 since a note says that a pupil read it to his teacher in Rajab/ September of that year].

Strachan's Catalogue No.: 53 National Library Naples: 87
Title: *Mu'allaqāt*
Author: Preislamic Arab poets with commentary by az-Zuawazanī
Scribe: Naṣrallāh ibn Ṣāliḥ
Strachan's comments: Septem poetae Arabes gentiles qui ante tempora Mohammedis floruerunt cum continuo Comentario Viri doctissimi Husein el Zuzeni, hunc librum propter raritatem et chari tatem venalem non repperit, ideoque describe ex antiquis mss. curavit Babilonis (sic) anno D. 1619 Georgius Strachanus Merniensis Scotus.

The seven Arabic heathen poets, who flourished before Mohammed's time, together with a continuous commentary by the very learned Husein el Zuzeni. This book was copied at Babylon in AD 1619 from some ancient manuscripts for George Strachan, of the Mearns Scot since he could not afford to buy it on account of its rarity and high price.
Dellavida's comments: The *Mu'allaqāt*, i.e. the seven poems of Preislamic Arabia, together with an ample commentary by az-Zuawazanī (d. 486/1093). The copyist who finished his work on 15 Ṣafar 1028/1 February 1619 was the same Naṣrallāh ibn Ṣāliḥ who copied Cat. No 64.

Strachan's Catalogue No.: 54 Vatican Library Arabic: 526
Titles:

1. *Kitāb al-Istidrāk*
2. (taqrīḍāt) appraisals written in favour of al-Ḥusayn ibn Ibrāhīm an-Naṭanzī
3. proverbs collected by the poet Abu'l-Faḍl 'Abdarraḥīm al-Mīkālī
4. proverbs by Ḥamza al-Iṣfahānī

Authors:

1. Abū Bekr az-Zubaydī
2. Unknown
3. Abu'l-Faḍl 'Abdarraḥīm al-Mīkālī
4. Ḥamza al-Iṣfahānī

Strachan's comments: In hoc libro haec continentur. Io Liber Aetymologiae nobilissimi Grammathici Sibawei cum declarationibus et additamentis Abubequir Andalutii. 2. Epigram(m)ata quoru(m)dam veteru(m) Gram(m)ichorum Arabum as logicam spectantia Et elequentiam. 3. Libellus proverbioru(m) ex alcorano, et alioru(m) Arabu(m) scriptis authore Abi Afadal il munkalj. Ordine Alfabetico. 4. Compendium proverbiorum Rariorum quae in libris antiquoru(m) Arabum reperiuntur, cum eorum explicationibus et causis. Liber Arabicantibus ultissimus. Emit Babilonj Anno Dnj 1619 Georgius Strachanus Merniensis Scotus.

In this book are the following works: 1. The Book of Etymology by the most famous grammarian Sibawei, with explanations and additions by Abubequir the Andalusian. 2. Epigrams by certain old grammarians, concerning Logic and Eloquence. 3. A pamphlet on proverbs taken from the Koran and other writings of the Arabs, by Ali Afadal il Munkaly in alphabetical order. 4. A summary of rare proverbs found in the books of the Arabs, with their explanation and origin. A book very useful to students of Arabic. **Dellavida's comments:** Strachan understood pretty well the contents of this manuscript, one of the most precious in his collection. It consists of several philological tracts of a high antiquity, which are very scarce, or even unique. 1. Is the *Kitāb al-Istidrāk*, 'The Book of Addition' by the Spanish grammarian Abū Bekr az-Zubaydī (d. 379/989), and contains additions to and corrections of the theory of nominal forms as expounded by Sībawayh, the founder of Arabic grammar. In his edition Guidi pointed out the value of this unique copy. 2. Is something slightly different from what Strachan thought it was: after two poems of unidentified authorship, we find a number of appraisals (*taqrīḍāt*) written in 458 and 459/1065–7 in favour of al-Ḥusayn ibn Ibrāhīm an-Naṭanzī (d. 497/1103 or 499/1106), a good scholar both in Arabic and Persian philology. The history of the bilingual territory of Western Persia during the 5th/11th century may gain some new light from the study of those still unpublished documents. 3. Is a collection of proverbs collected by the poet Abu'l-Faḍl 'Abdarraḥīm al-Mīkālī (d. 436/1044) which Strachan described correctly although he misread the author's name. 4. Was omitted by Strachan and is of dubious identification . . . Finally, 5. (Strachan's 4) is an abbreviation of a work by Ḥamza al-Iṣfahānī (died before 360/970) on proverbs set up in the form of a comparison.

This very old manuscript dated 622–3/1225–6 was written by Muḥammad ibn Ḥiṣn ibn Ḥamdūn al-Wāsiṭ, who was a scholar himself and accomplished his work with painstaking care.

Strachan's Catalogue No.: 55 Vatican Library Arabic: 488
Titles:

1. *Laṭā'if al-ishārāt fī asrār al-falak wa'l-ḥurūf al-ma'nawiyyāt*
2. *Kashf al-asrār 'ammā khafiya 'an al-afkār*

Authors:

1. al-Būnī
2. al-Afqahsī

Strachan's comments: Hic habentur 1o Liber cui nomen shems ilmaaref, i. sol Intelligibilium. Opus cabalisticu(m) (secundu(m) opini(on)um pithagorae) o(mn)ia in numeris et literis ponens. 2o Liber questionu(m) cu(m) suis responsionibus supra Alcoranu(m). Authore Gelal Iddin

Iddauani. Emit Babiloni anno D. 1619 Georgius Strachanus Merniensis Scotus.

Here are 1. A book bearing the title shems ilmaaref, i.e. The sun of intelligible things. A cabalistic work, according to the doctrine of Pythagoras, setting everything up into numbers and letters. 2. A book of queries and answers on the Koran, by Gelal Iddin Iddauani.

Dellavida's comments: Al-Būnī (d. 622/1225), the most famous Arabic writer on magic, actually is the author of a work entitled *Shams al-ma'ārif wa-laṭa'if al-'awārif*, 'The sun of knowledge and kindness of benefits', the classic text book on this subject. Our manuscript, however, contains another of his works, *Laṭā'if al-ishārāt fī asrār al-falak wa'l-ḥurūf al-ma'nawiyyāt*, 'The subtle indications concerning the secrets of the heavenly spheres and the esoteric letters'. Strachan omitted the mention of two anonymous works on magic which follow al-Būni's book; he thought, probably, that they were two appendices of it. The last work in the manuscript bears a false title, which Strachan faithfully reproduced: *al-As'ila ash-sharīfa al-qur'āniyya*, by ad-Dawwānī. Among his works, however, none bears this title. Actually, the manuscript contains a work by al-Afqahsī (d. 808/1405) dealing with difficult and subtle questions of theology and law, entitled *Kashf al-asrār 'ammā khafiya 'an al-afkār*, 'The Unveiling of the secrets which are concealed to intelligence'.

There is no date in the manuscript, which is contemporary with or a little earlier than Strachan's time.

Strachan's Catalogue No.: 57 Vatican Library Arabic: 476
Titles:

1. *Risālat aṭ-ṭayf*
2. Miscellaneous poems
3. *Dā'irat an-nujūm*
4. *Lam'at ash-shākī wa-dam 'at al-bākī*
5. *al-Jawhar al-fard fi'l-munāẓara bayna an-narjis wa'l-ward*

Authors:

1. Bahā'addīn al-Irbilī
2. Anonymous poets
3. Sibṭ al-Māridīnī
4. Ṣalāḥaddīn aṣ- Ṣafadī
5. Ibn Musharraf al-Māridīnī

Strachan's comments: In hoc libro haec continentur 1o Opusculum cuj nomen Somnium. Autore Baha iddin f. Isa. Arbellano Medo. Opus eloquij et humanitatis plenum. 2o poemata diversa variorum(m) poetarum in laudem eiusdem Althoris (sic). 3o Opusculum, cui titulus Revolutio stellarum authore M. Il Mardini. 4. Dialogus stilo politissimo, amorus plenus, cui nomen est Incendium Dolentis et Lacrimae plorantis. A(uthore) Gemal Iddin 5o Discursus continens disputationem seu contentionem Inter Violam et Rosam opusculu(m) elegans et subtile Ibn Mishref Mardinj. Emit Babiloni Anno Dnj 1619 Georgius Strachanus Merniensis Scotus.

In this book are contained: 1. A pamphlet entitled the Dream by Bahaiddin son of Isa the Arbelian and Median, a work full of elegance and grace. 2. Diverse poems by various poets, in praise of that author. 3. A pamphlet entitled the Revolution of the Stars, by M. Il Mardini. 4. A dialogue in the most polished style and filled with love, entitled The Fire of the complainer and the Tears of the weeper. By Gemal Iddin. 5. A discourse containing a discussion or an argument between the Carnation and the Rose, an elegant and subtle pamphlet by Ibn Mishref Mardinj.

Dellavida's comments: The most significant among the works contained in this miscellaneous manuscript have been correctly described by Strachan. 1. Is *Risālat aṭ-ṭayf*, a romantic story of a love dream, which might well support the views of those who maintain that the Provençal and Old Italian theory of love is of Arabic origin. Its author, Bahā'addīn al-Irbilī (i.e. from Irbil, Arbela of the classical authors), died in 692/1293. The poems and fragments of poems immediately subsequent do not refer, as Strachan assumed, to Bahā'addīn. 3. Is *Dā'irat an-nujūm*, 'The circle of the stars' by the well known Shī'ite astronomer Sibṭ al-Māridīnī (d. 912/1506). 4. Is an essay on homosexual love by the Egyptian author, Ṣalāḥaddīn aṣ- Ṣafadī (d. 764/1363) and its Arabic title is *Lam'at ash-shākī wa-dam 'at al-bākī*. Finally, the contrast between the narcissus and the rose (*al-Jawhar al-fard fi'l-munāẓara bayna an-narjis wa'l-ward*) belongs to a literary genus much appreciated among the Arabs, is a work of Ibn Musharraf al-Māridīnī (d. about 850/1446–7).

The handwriting of the manuscript is Persian, and it may belong to the second half of the 9th/15th century. However, it must have been brought to Syria, because one of the names of its owners, appearing on the title page, is that of Abdarraḥmān ibn Abī Bekr ibn Muḥammad al-'Aynī, a man known in literary history, who spent most of his life in Damascus and who died there in 893/1488. The manuscript later went through several hands; in the month of Shawwāl 1027/September-October 1618, a year before Strachan acquired it, it had been read by Omar ibn Aḥmad al-Ḥanafi al-Qādirī.

Strachan's Catalogue No.: 59 National Library Naples: III.F.31
Titles:

1. *Risāla-I Mu'īniyya*
2. *Risāla*
3. *ar-Risāla ash-Shamsiyya fi'l ḥisāb*

Authors:

1. Not recorded
2. Maḥmūd ibn Muḥammad ibn Quwām al-Wāshtānī
3. Niẓāmaddin an-Naysābūrī

Strachan's comments: Sphera et opuscula duo De Arithmetica persica lingua conscripta. Emit Babilonj anno Dnj 1619 Georgius Strachanus Merniensis Scotus.

The Sphere and two pamphlets on Arithmetic written in the Persian language.

Dellavida's comments: This is the only Persian item among the Strachan manuscripts at Naples. The tract which Strachan called 'Sphere' is the *Risāla-I Mu'īniyya*, 'The Helping Tract' or 'Tract dedicated to Mu'īnaddīn or Mu'īnaddawala', a classical summary of astronomy by Naṣīraddīn Ṭūsī' (i.e. native of the Persian town Ṭūs' or Ṭōs, one of the greatest Muslim astronomers and mathematicians (597/1201-672/1274) and founder and director, by order of the Tatar khan Hūlāghū, of a famous observatory at Marāgha in Ādherbījān, remnants of which are still extant. He also wrote extensively on law, medicine and Shī'te theology.

The second tract is a summary of arithmetic generally known as *Risāla* by the Qāḍī Maḥmūd ibn Muḥammad ibn Quwām al-Wāshtānī . . . Also the third tract is a short treatise on arithmetic; it bears no title . . . [but] in all likelihood is the Persian translation of *ar-Risāla ash-Shamsiyya fi'l ḥisāb*, 'The Solar Tract on Arithmetic', by Niẓāmaddin an-Naysābūrī, a scholar of the 9th/14th century.

Still in original binding of reddish-brown leather.

Strachan's Catalogue No.: 61 Vatican Library Arabic: 569
Title: *Tafsīr al-manāmāt*
Author: Muḥammad ibn Sīrīn
Strachan's comments: Explicatio insomniorum auth(ore) Ben Sirin. Emit Babiloni an(n)o D. 1619 G. Strachanus Merniensis Scotus.
Explanation of dreams by Ben Sirin.
Dellavida's comments: This manuscript, of recent date and devoid of any particular significance, contains one of the many versions in which the pseudepigraphic Book of dreams (*Tafsīr al-manāmāt*) by Muḥammad ibn Sīrīn has been preserved. The author is said to have lived in the first century after the Hijra and to have been of Aramaic origin. Actually the work is chiefly based on Greek sources. It was translated into Latin and was widely known in Europe during the Middle Ages.

Strachan's Catalogue No.: 63 Vatican Library Arabic: 510
Untitled but contains verses from *Dīwān aṣ-ṣabāba*
Author: Ibn Abī Ḥajala (d. 776/1375)

Strachan made no comments on this manuscript but the catalogue no. entry is in his handwriting.
Dellavida's comments: On the flyleaf near the entry 'No. 8' the Hebrew word 'Berākhāh', 'Blessing' is written. The same word with the wrong spelling 'Abrākhāh' is found on Vat. Arabic: 510, which bears the entry 'No. 63'. This manuscript, which is not dated, may belong to the end of the 10th/16th or the beginning of the 11th/17th century and is of undoubted Persian origin, as appears from its paper and binding, and from several Persian verses scribbled on its flyleaf. It contains a well-known collection of poems and stories referring to famous lovers, the *Dīwān aṣ-ṣabāba* by Ibn Abī Ḥajala (d. 776/1375).

Strachan's Catalogue No.: 64 Vatican Library Arabic: 558
Title: *Saqṭaz-zand*
Author: Abdu'l-'Alā al-Ma'arrī
Strachan's comments: Favilla ferri ignarii. Opus poeticum ingegnosissim(um) et elequentissim(um). In quo observantur o(mn)es figurae stili poetici, versuu(m)que Araborum differentiae. Auth(ore) Abi Ileela il Muarrj. Describi ex antiquiss(i)mo exemplari curavit Babilloni an(n)o Dni 1619 Georgius Strachanus Merniensis Scot.

The Spark of the Fire Steel. A most ingenuous and eloquent book of verse, in which all figures of poetical style and all varieties of Arabic verse are observed, by Ali ileela il Muarrj. It was copied at Babylon in AD 1619 from a very ancient manuscript for George Strachan of the Mearns Scot.
Dellavida's comments: Strachan's description fits pretty well the extremely elaborate style of the *Saqṭaz-zand*, a collection of poems by the blind poet Abdu'l-'Alā al-Ma'arrī (d. 449/1057), one of the most prominent personalities in Arabic literature.

The copyist of the manuscript was Naṣrallāh ibn Ṣāliḥ, who gives himself and his father the title of a 'Sayiyd' i.e. a descendant from the caliph 'Alī and his wife Fāṭima, the Prophet's daughter. Besides this manuscript, dated Rabī' II 1028/March-April 1619, Naṣrallāh copied another manuscript for Strachan (Strachan Cat. No. 53). Still in original binding of black leather with gold stamping.

Mss Uncatalogued by Strachan

Ms. 1 of 6 Vatican Library Arabic: 585
Title: *Talkhīṣ al-Miftāḥ*
Author: al-Qazwīnī Khaṭīb Dimashq
Strachan's comments: Haec est Rhetorica Arabica, cui nomen est Explicatio Clavis. Emit Alepi anno Dni. 1615. Georgius Strachanus Merniensis Scotus. This is an Arabic Rhetoric entitled Explanation of the Key.
Dellavida's comments: The *Talkhīṣ al-Miftāḥ* by al-Qazwīnī Khaṭīb Dimashq (d. 739/1338) is a summary (this is the meaning of '*talkhīṣ*'

rather than explanation) of the third part of *Miftāḥ as-sa'āda* 'The Key of Happiness', a treatise on grammar and rhetoric by as-Sakkāki (d. 626/1229).

McInally's comment: This appears to be one of the manuscripts that Strachan had left in Aleppo and that Fr Vincent retrieved on his way home to Rome.

Ms. 2 of 6 National Library Naples: 22

Title: *Tafsīr al-Jalālayn*

Author: Jalāladdin al-Maḥallī and Jalāladdin as-Suyūṭī

Strachan's comments: Explicatio literalis alcorani per duos Gelalos. Arabes tafsir el Gelalein vocant hunc librum quem ex manuscripto antiquo describe curavit Babilonij anno Dnj 1619 Georgius Strachanus Merniensis Scotus.

Literal explanation of the Koran by the two Gelals. It was transcribed at Babylon in 1619 from an old manuscript for George Strachan of the Mearns Scot.

Dellavida's comments: *Tafsīr al-Jalālay*, a very popular and elementary commentary on the Koran, begun by Jalāladdin al-Maḥallī (d. 844/1459) and achieved by Jalāladdin as-Suyūṭī (d. 911/1505). The copy was executed in the months of Muḥarram and Ṣafar 1028/December 1618 to February 1619 by Ibrāhīm ibn Ḥasan, who also copied Cat. Nos 41 and 43.

Ms. 3 of 6 National Library Naples: 38

Title: *Minhāj al-karāma fi ma'rifat al-imāma*

Author: Ibn al-Muṭahhar al-Ḥillī

Strachan's comments: Liber apologeticus de Jure successionis Aly ibn Abytaleb post mortem Mahometis Jllumque cum suis successoribus esse praelatos legittimos (sic) contra sectam traditiuam, qui asserunt Abubekir Omar et Othman Illi legittimo (sic) successisse. Hic liber apud persas magni estimator. Emit Babilonj anno Dnj 1617 Georgius Strachanus Merniensis Scotus.

An apologetic book on the right of succession of Aly ibn Abytaleb after Mahomet's death, maintaining that he and his successors are the legitimate High Priests, as against the sect of the Traditionalists, who maintain that Abubekir, Omar and Othman were his [i.e. Mohammad's] legitimate successors. This book is highly appreciated among the Persians.

Dellavida's comments: Strachan is quite correct in his statements. The book known as *Minhāj al-karāma fi ma'rifat al-imāma*, 'The Path of Grace concerning the Knowledge of the Caliphate' is a treatise on the Shī'te theory of the caliphate, by Ibn al-Muṭahhar al-Ḥillī (see Strachan Cat. No. 44). As is well known, the Shī'a contention is that the right to the caliphate belongs to the family of 'Alī, and that the Omayyad and Abbasid caliphs who were the actual rulers of Islam after 'Ali's death were nothing but usurpers. The Shī'ite tenets became the official religion of Persia in the sixteenth century, as against the beliefs of the 'Traditionalists' or Sunnī.

This manuscript was written at the beginning of Dhu'lqa'da 1005/June 1597.

Ms. 4 of 6 Majlis Shūrā Library, Tehran: Ms. 2060
Title: *Taḥrīr Uqlīdis – Euclid's Elements* plus two commentaries
Author: Translation and first commentary Thābit ibn Qurra. Second commentary al-Ḥajjāj.
Scribe: Describes himself as 'aṣgar (?) al-'ibād', 'least of the servants', but no name is given.
Strachan's comments: Elementa Euclidis ex graecis in Arabicum sermonem (?) traducta a Thabet Ibn Curra Harran Mesapotamia cum Commentario eiusdem authoris in eadem elementa. Item Commentarius in eadem ab Hagiagis Arabo Doctissimo. Georgius Strachanus Merniensis Scotus.

Translation from Greek to Arabic of Euclid's Elements by Thabet Ibn Curra Harran of Mesapotamia with a commentary by the same author. Also a commentary of the same work by the learned Arab Hagiagis.
De Young's comments: Strachan's note is an expansion of the Arab title *Taḥrīr Uqlīdis* of Thābit ibn Qurra and also the commentary of al-Ḥajjāj on the *Taḥrīr Uqlīdis* . . . [T]here is written in red ink 'Spahan 1624' and below it another line which appears to be its price 'Ab. 15-'.

Ms. 5 of 6 British Library Additions: 7720
Titles:

1. *ar-Risāla al-kubrá fi'l manṭiq*
2. *Jām-i gītī-numā*

Authors:

1. 'Alī ibn Muḥammad al-Jurjānī
2. Qādī Mīr Ḥusayn ibn Mu'īnaddīn al-Maybudī

Strachan's comments: Universum, seu, ut Persae vocant, poculum mundi, opera Georgii Strachani Merniensis Scoti in Latinum idoma traducta (sic), 1634.

The Universe or, as the Persians call it, The World Cup, translated into Latin by George Strachan of the Mearns Scot, 1634.
Dellavida's comments: A Persian manuscript . . . contains a treatise on logic by the 14th century author 'Alī ibn Muḥammad al-Jurjānī and a book on popular philosophy, the *Jām Gētī Numā* 'The World Mirror'. Both works are accompanied by a Latin interlinear translation and *Jām Gētī Numā* is accompanied by [Strachan's] endorsement. [Claude J. Rich purchased this manuscript in Baghdad in second decade of the 19th century.]
Pourjavady's comments: It is evident from this copy that Strachan was familiar with Ibn Sīna. He renders *Shaykh al-ra'īs* as 'first doctor' and offers

the further explanation 'i.e. Avicenna' . . . In some of his marginal notes, Strachan reveals his disagreement with the argument and betrays his religious disquiet with the contents of the work . . . Strachan's critical study of this treatise on Islamic philosophy in its original Persian language represents the first example of such an endeavour in the early modern period.

Ms. 6 of 6 Cambridge University Library No.: XIX Additions 252
Title: *Tarjumān-i-Qur'ān 'The Interpreter of the Qur'ān'*
This manuscript is in Persian and was completed on 20 Jumādá I, 1033/10 March 1624 for George Strachan in Isfahan. At a later date it was gifted to a Rev. George Lewis.

Archives

Archivum Romanum Societatis Iesu (ARSI)
  ARSI, *Anglia*, 42
Archivum Secretum Vaticanum (ASV)
Biblioteca Nazionale Vittorio Emanuele III (Naples)
Bibliotheca Apostolica Vaticana (BAV)
  Ms. Barb. Lat. 2081
  Ms. Barb. Lat. 2190
Bibliothèque Nationale (Paris) (BN)
Majlis Shūrāh Library, Tehran
  Ms. 2060
Scottish Catholic Archives (SCA), University of Aberdeen, Special Collections
  CB/57/12 (George Strachan's *Album Amicorum*)

Bibliography

Primary Sources

Albèri, E. (1839), *Relazioni degli ambasciatori veneti al Senato,* Florence.

Anderson, P. J. (ed.) (1893), *Fasti Academia, Officers and Graduates of University and King's College,* Aberdeen.

Anderson, P. J. (ed.) (1906), *Records of the Scots Colleges,* vol. 1, Aberdeen.

Balfour, Sir J. (ed.) (1904), *The Scots Peerage,* vol. 1, Edinburgh.

Beg, U. (1604), *Relaçiones de Don Juan de Persia,* Madrid.

Bender, J. (ed.) (1868), *Geschichte Der Philosophischen und Theologischen Studien in Ermland,* Braunsberg.

Bruce, J. (ed.) (1868), *Journal of a Voyage into the Mediterranean: Edited Extracts from Bibliotheca Digbeiana, 1680,* London.

Burke, J. (1884), *The General Armory of England, Scotland, Ireland and Wales; Comprising a Registry of Armorial Bearings from the Earliest to the Present Time,* London.

Calendar of State Papers (1857), London.

Calendar of State Papers (1870), *Colonial, East Indies, 1617–1621,* London.

Chick, H. (ed. and trans.) (1939), *A Chronicle of the Carmelites in Persia and the Papal Mission of the XVII and XVIII centuries,* London.

Della Valle, P. (1664), *Viaggi di Pietro Della Valle il Pellegrino . . .,* Venice.

Dellon, C. (1687), *L'Inquisition de Goa. La relation de Charles Dellon,* Paris.

Dempster, T. of Muiresk (1627), *Historia Ecclesiastica Gentis Scotorum,* Bologna.

Dempster Thomas (1829), *Historia Ecclesiastica Gentis Scotorum sive De Scriptoribus Scotis,* Edinburgh.

Desideri, I., S. J. (1728), *Difesa della Compagnia di Giesu in ordine alla Missione del Tibet,* Rome. Referenced in Wessels, B. (1924), *Early Jesuit Travellers,* The Hague, p. 83.

Erpenius, T. (1620), *Oration on the Value of the Arabic Language,* Leiden.

Erpenius, T. (1620), *Rudimenta Linguae Arabicae,* Leiden.

Franco, A., S. J. (1717), *Imagem da Virtude em o Noviciado da Companhia de Jesu na Corte de Lisboa,* Coimbra. On pp. 400–3 Franco quotes *Lettre annue del Tibet del MDCXXVI et della Cina del MDCXXIV scritte al M. R. P. Mutio Vitelleschi, Generale della Compagnia di Giesù* (Rome, 1628).

De Silva y Figueroa, G. (1667), *L'Ambassade de Don Garcías de Silva Figueroa en Perse,* Paris.

Hebermann, C. G. et al. (eds) (1913), *The Catholic Encyclopedia,* New York.

Leask, W. K. (1910), *Musa Latina Aberdonensis*, Aberdeen, pp. 338–46.

Lithgow, W. (1632), *The Total Discourse of the Rare Adventures and Painefull Peregrinations of long nineteen Years Travayles from Scotland to the most Famous Kingdomes in Europ, Asia and Affrica*, London.

Masson, D. (ed.) (1884), *Register of the Privy Council of Scotland (RPCS)*, vol. 5 (1592–9), vol. 6 (1599–1604), vol. 7 (1604–7), Edinburgh.

Nicolay, N. de (1576), *Les Navigations, Peregrinations et Voyages faictes en la Turquie*, Anvers.

O'Neill, C. E. and Domínguez, J. M. (eds) (2001), *Diccionario Historico de la Compañia de Jesus, IHSI*, vol. II, Rome.

Rogers, C. (1873), *Estimate of the Scottish Nobility during the Reign of James VI*, London.

Sainsbury, W. N. (ed.) (1870), *Calendar of State Papers, Colonial Series, East India, China and Japan*, 1617–1621, London (Court Minutes of the East India Company, No. 535, 8 January 1619).

Sandys, G. (1621), *A Relation of a Journey begun AD 1610*, 2nd edn, London.

Santo Brasca (1481), *Viaggio alla sanctissima cita di Ierusalem*, Milan.

Skinner, J. (1624), *A True Relation of the unjust, cruell, and barbarous proceedings against the English at Amboyna*, London.

Soimonov, F. I., *Auszug aus dem Tage-Buch . . . Soimonov*, in Müller, G. F. (1762), *Sammlung russischen Geschichte VII*, St Petersburg.

Strachan, G. (1607), *Luciani Samosatensis Insignis Oratio de non temere credendo calumniate Recensuit et Latinam fecit* Georgius Strachanus Merniensis Scotus, Paris.

Valerio, A. (1679), *Historia della Guerra di Candia di Andrea Valerio Senatore Veneto*, Venice.

Voltaire (1775), *Oeuvres de Mr de Voltaire, Tome Vingtieme, Histoire du Parlement de Paris*, Paris.

Secondary Sources

Agoston, G. (2007), 'Information, Ideology and Limits of Imperial Policy: Ottoman Grand Strategy in the Context of Ottoman-Habsburg Rivalry', in Aksan, V. H. and Goffman, D. (eds), *The Early Modern Ottomans: Remapping the Empire*, Cambridge, pp. 75–103.

Allen, C. (2000), *The Search for Shangri-La: A Journey into Tibetan History*, London.

Ames, G. (2012), 'Acts of Faith and State: The Goa Inquisition and the French Challenge to the Estado da India', in Elbl, I. (ed.), *Portuguese Studies Review* 17(1) (Collected Essays in Memory of Glenn J. Ames): 11–35.

Ammannati, F. (2018), 'Book Prices and Monetary Issues in Renaissance Europe', in Granata, G. and Nuovo, A. (eds), *Printed Book Sale Catalogues and Private Libraries in Early Modern Europe*, Macerata, pp. 161–77.

Aranha, P. (2014), 'The Social and Physical Spaces of the Malabar Rites

Controversy', in Marcocci, G., Pavan, I. and de Boer, W. (eds), *Space and Conversion in Global Perspective*, Leiden, pp. 214–32.

Ari, B. (2004), 'Early Ottoman Diplomacy: Ad Hoc Period', in Yurdusev, A. N. (ed.), *Ottoman Diplomacy: Conventional or Unconventional?*, Basingstoke, pp. 36–65.

Beretta, M. (2009), 'The Revival of Lucretian Atomism and Contagious Diseases during the Renaissance', *Medicina nei Secoli* 15(2): 129–54.

Bernardini, M. (2011), 'Giovan Battista and Gerslamo Vecchietti in Hormuz', in Matthee, R. and Flores, J. (eds), *Portugal, the Persian Gulf and Safavid Persia*, Leuven, pp. 265–81.

Black, J. B. (1959), *The Reign of Elizabeth 1558–1603*, Oxford.

Blow, D. (2009), *Shah Abbas: The Ruthless King who Became an Iranian Legend*, London.

Bosworth, C. E. (2007), *An Intrepid Scot: William Lithgow of Lanark's Travels in the Ottoman Lands, North Africa and Central Europe 1609–21*, Farnham.

Bosworth, C. E. (2011), 'Three British Travellers to the Middle East and India in the Early Seventeenth Century', *Graeco Arabica* XI: 187–97.

Bosworth, C. E. (2012), *Eastward Ho, Diplomats, Travellers and Interpreters of the Middle East and Beyond 1600–1940*, London, chapter 2.

Bournoutian, G. A. (2002), *A Concise History of the Armenian People (From Ancient Times to the Present)*, Ann Arbor, MI.

Bousama, W. J. (1968), *Venice and the Defense of Republican Liberty: Renaissance Values in the Age of the Counter Reformation*, Berkeley, CA.

Braudel, F. (1972–3), *The Mediterranean and the Mediterranean World in the Age of Philip II* (trans. Sian Reynolds), 2 vols, Swindon.

Browne, E. G. (1896), *Catalogue of Persian Manuscripts in the University of Cambridge*, Cambridge.

Bull, G. (trans.) (1989), *The Journals of Pietro Della Valle, The Pilgrim*, London.

Burman, T. E. (1998), 'Tafsīr and Translation: Traditional Arabic Qur'ān Exegesis and the Latin Qur'āns of Robert of Ketton and Mark of Toledo', *Speculum* 73: 703–32.

Carey, D. (2009), *Continental Travel and Journeys beyond Europe in the Early Modern Period: An Overlooked Connection*, London.

Chaudhuri, K. N. (1965), *The English East India Company: The Study of an Early Joint-Stock Company 1600–1640*, London.

Coleman, A. P. (2004), 'Ochterlony, Sir David (1758–1825)', *Oxford Dictionary of National Biography*, Oxford.

Coleridge, Rev. H. J. (1997), *The Life and Letters of Francis Xavier 1506–1556*, New Delhi.

Costa, Fr C. J. (2006), *St Paul's College and Rachol Seminary*, Goa.

Dalrymple, W. (1990), *In Xanadu a Quest*, London.

Dellavida, G. L. (1956), *George Strachan: Memorials of a Wandering Scottish Scholar of the Seventeenth Century*, Aberdeen.

De Young, G. (2015), 'Two Hitherto Unknown Arabic Euclid Manuscripts', *Historia Mathematica* 42: 132–54.

Dilworth, M. (2004), 'Chisholm, William (1547–1629)', *Oxford Dictionary of National Biography*, Oxford.

Doubleday, S. R. (2015), *The Wise King: A Christian Prince, Muslim Spain, and the Birth of the Renaissance*, New York.

Durkan, J. (1971), 'Notes on Scots in Italy', *Innes Review* 22(1): 12–18.

Dursteler, E. (2002), 'Commerce and Coexistence: Veneto-Ottoman Trade in the Early Modern Era', *Turcica* 34: 105–33.

Dursteler, E. (2006), *Venetians in Constantinople Nation, Identity, and Coexistence in the Early Modern Mediterranean*, Baltimore, MD.

Dursteler, E. R. (2009), 'Power and Information: The Venetian Postal System in the Early Modern Eastern Mediterranean', in Dursteler, E. R., Curto, D., Kirshner, J. and Testo, F. (eds), *From Florence to the Mediterranean: Studies in Honor of Anthony Molho*, Florence, pp. 601–23.

Dursteler, E. R. (2012), 'Speaking in Tongues: Language and Communication in the Early Modern Mediterranean', *Past and Present* 217(1): 47–77.

Du Toit, A. (2004), 'Thomas Dempster (1579–1625)', *Oxford Dictionary of National Biography*, Oxford.

Elger, R. (1984), *Muṣṭafā al-Bakrī*, Schenefeld.

Elgood, C. (2010), *A Medical History of Persia and the Eastern Caliphate*, Cambridge.

Endress, G. (2006), 'Reading Avicenna in the Madrasa: Intellectual Genealogies and Chains of Transmission of Philosophy and the Sciences in the Islamic East', in Montgomery, J. E. (ed.), *Arabic Theology, Arabic Philosophy: From the Many to the One: Essays in Celebration of Richard M. Frank*, Leuven, pp. 371–422.

Faroqhi, S., McGowan, B. and Pamuk, S. (eds) (1997), *An Economic and Social History of the Ottoman Empire*, vol. 2, Cambridge.

Fischer, T. A. (1902), *The Scots in Germany*, Edinburgh.

Fitzpatrick, E. A. (ed.) (1933), *St Ignatius and the Ratio Studiorum*, New York.

Flannery, J. (2013), *The Mission of the Portuguese Augustinians to Persia and Beyond (1602–1747)*, Leiden.

Forbes-Leith, W. (1885), *Narratives of Scottish Catholics under Mary Stuart and James VI*, London.

Forbes-Leith, W. (1909), *Memoirs of Scottish Catholics during XVIIth and XVIIIth Centuries*, London.

Forster, W. (1906), *The English Factories in India 1622–1623*, Oxford.

Fück, J. (1955), *Die Arabischen Studien in Europa bis in den Anfang des 20. Jahrhunderts*, Leipzig.

Furber, H. (2004), 'Rival Empires of Trade 1600–1800', in McPherson, K. (ed.), *Maritime India*, New Delhi, pp. 163–293.

Glei, R. F. and Tottoli, R. (2016), *Ludovico Marracci at Work: The Evolution of his Latin Translation of the Qur'ān in Light of his Newly Discovered Manuscripts*, Wiesbaden.

Golubovich, P. G. (1906), *Biblioteca Bio-Bibliografica della Terra Sancta e dell'Oriente Francescana*, Florence.

Gordon, J. F. S. (1869), *Catholic Church in Scotland*, Glasgow.

Gordon, T. J. (2013), *Renaissance Emir: a Druze Warlord at the Court of the Medici*, London.

Goyeau, G. (1910), 'Family of Harlay', in Heberman C. G., Pace, E., Pallen, C., Shahan, T. and Wynne, J. (eds), *The Catholic Encyclopedia*, New York.

Gribben, C. and Mullan, D. G. (2009), *Literature and the Scottish Reformation*, Farnham.

Gründler, B. (2016), 'Aspects of Craft in the Arabic Book Revolution', in Renn, J. and Bentjes, S. (eds), *Globalization of Knowledge in the Post-Antique Mediterranean 700–1500*, London, pp. 31–66.

Guiley, R. (2008), *The Encyclopedia of Witches, Witchcraft and Wicca*, New York.

Gulbenkian, R. (1981), *The Translation of the Four Gospels into Persian*, Immensee.

Gürkan, E. S. (2015), 'Mediating Boundaries: Mediterranean Go-Betweens and Cross-Confessional Diplomacy in Constantinople, 1560–1600', *Journal of Early Modern History* 19(2–3): 107–28.

Halft, D. (2017), 'The Arabic Vulgate in Safavid Persia', unpublished PhD dissertation, Freie Universität Berlin.

Halloran, B. (2003), *The Scots College Paris 1603–1792*, Edinburgh.

Hamilton, A. (2001), 'The Study of Islam in Early Modern Europe', *Archiv für Religionsgeschichte* 3(1): 169–82.

Harper, J. G. (2011), 'Introduction', in Harper, J. G. (ed.), *The Turk and Islam in the Western Eye, 1450–1750*, Burlington, VT, pp. 1–20.

Haynes, J. (1986), *The Humanist as a Traveller: George Sandys's Relation of a Journey Begun An: Dom: 1610*, Rutherford, NJ.

Hayes, K. J. (2004), 'How Thomas Jefferson read the Qur'ān', *Early American Literature* 39(2): 247–61.

Heberman C. G., Pace, E., Pallen, C., Shahan, T. and Wynne, J. (eds) (1905–14), *Catholic Encyclopaedia*, New York.

Hirschler, K. (2016), *Medieval Damascus: Plurality and Diversity in an Arabic Library*, Edinburgh.

Ikram, S. M. (1964), *Muslim Civilization in India*, New York.

Jacoby, D. (2004), 'Silk Economics and Cross-Cultural Artistic Interaction: Byzantium, the Muslim World and the Christian West', *Dumbarton Oaks Papers* 58: 197–240.

Karamustafa, A. T. (1994), *God's Unruly Friends: Dervish Groups in the Islamic Later Middle Period, 1200–1550*, Salt Lake City.

Klein, D. (2007), *Die osmanischen Ulema des 17. Jahrhunderts: Eine geschlossene Gesellshaft?*, Berlin.

Knobloch, E. (2002), 'La connaissance des mathématiques arabes par Clavius', *Arabic Sciences and Philosophy* 12(2): 257–84.

Küçükhüseyin, Ş. (2017), 'Messianic Expectations among Anatolian Turkmen', in Korn, L. and Müller Wiener, M. (eds), *Central Periphery?*

Art, Culture and History of the Medieval Jazira (Northern Mesopotamia, 8th–15th Centuries), Wiesbaden, pp. 229–38.

Kurt, A. (2013), 'The Search for Prester John, a Projected Crusade and the Eroding Prestige of the Ethiopian Kings, c. 1200–1540', *Journal of Medieval History* 39(3): 297–320.

Lefranc, A. (1893), *Histoire du Collège de France*, Paris.

Lockhart, L. (1958), *The Fall of the Safavid Dynasty and the Afghan Occupation of Persia*, Cambridge.

Loop, J. (2017), 'Arabic Poetry as Teaching Material in Early Modern Grammars and Textbooks', in Loop, J., Hamilton, A. and Burnett, C. (eds), *The Teaching and Learning of Arabic in Early Modern Europe*, Leiden.

Lorenzen, D. N. (2005), *Who Invented Hinduism? Essays on Religion in History*, New Delhi.

Lucchetta, F. (1989), 'La scuola dei "giovani di lingua" veneti nei secoli XVI e XVII', *Quaderni di studi arabi* 7: 19–40.

McInally, T. (2012a), 'Scholars and Spies – Three Humanists in the Service of James VI/I', *Recusant History* 31(2): 135–46.

McInally, T. (2012b), *The Sixth Scottish University The Scots Colleges Abroad: 1575 to 1799*, Leiden.

Maclagan, Sir Edward (1932), *The Jesuits and the Great Mogul*, London.

MacLean, G. and Matar, N. (2011), *Britain and the Islamic World, 1558–1713*, Oxford.

McRoberts, Fr D. (1952), 'George Strachan of the Mearns, An Early Scottish Orientalist', *Innes Review* 3(2): 110–28.

Mainoni, P. (2000), 'La Seta in Italia fra XII e XIII secolo', in Mola, L., Mueller, R. and Zanier, C. (eds), *La Seta in Italia dal Medioevo al Seicento*, Venice, pp. 365–98.

Marco, G. (2018), 'Setting up a Silk Manufacture in the Late Middle Ages: Strategies and Outcomes in Comparison with Tuscany and the Crown of Aragon', *eHumanista/IVITRA* 14: 110–21.

Martin, F. X. (1992), *Friar, Reformer and Renaissance Scholar: Life and Work of Giles of Viterbo, 1469–1532*, Villanova, PA.

Matteoni, F. (2009), 'Blood Beliefs in Early Modern Europe' unpublished PhD dissertation, University of Hertfordshire.

Matthee, R. (1999), *The Politics of Trade in Safavid Iran: Silk for Silver, 1600–1730*, Cambridge.

Matthee, R. (2012), *Persia in Crisis: Safavid Decline and the Fall of Isfahan*, London.

Matthee, R. (2018), '*Zar-o Zur*: Gold and Force: Safavid Iran as a Tributary Empire', in Tomohiko, U. (ed.), *Comparing Modern Empires: Imperial Rule and Decolonization in the Changing World Order*, Sapporo, pp. 35–64.

Mills, S. (2017), 'Learning Arabic in the Overseas Factories: The Case of the English', in Loop, J., Hamilton, A. and Burnett, C. (eds), *The Teaching and Learning of Arabic in Early Modern Europe*, Leiden.

Montefiore, S. S. (2011), *Jerusalem the Biography*, London.

Moshenska, J. (2016), *A Stain in the Blood: The Remarkable Voyage of Sir Kenelm Digby*, London.

Mukherjee, R. (1974), *The Rise and Fall of the East India Company*, New York.

Müller, R. A. (1996), 'Student Education, Student Life', in de Ridder Simmons, H. (ed.), *A History of the University in Europe*, vol. 2, Cambridge, pp. 326–54.

Nahli, O., Frontini, F. and Monachini, M. (2016), 'Al Qamus al Muhit, a Medieval Arabic Lexicon in LMF', *LREC Conference Papers*, Paris, pp. 943–50.

Nasr, S. H. (2006), *Islamic Philosophy from its Origins to the Present: Philosophy in the Land of Prophesy*, New York.

Nicholl, C. (1999), 'Field of Bones: the Last Journey of Thomas Coryate, the English Fakir and Legstretcher', *London Review of Books* 21(17): 1–7.

Nickson, M. A. E. (1970), *Early Autograph Albums in the British Museum*, London.

Nisan, M. (2002), *Minorities in the Middle East: A History of Struggle and Self Expression*, Jefferson, NC.

Norman, D. (1960), *Islam and the West: The Making of an Image*, Edinburgh.

Ó Gráda, C. (2007), 'Making Famine History', *Journal of Economic Literature* 45(1): 5–38.

Onley, J. (2014), 'Indian Communities in the Persian Gulf, c. 1500–1947', in Potter, L. G. (ed.), *The Persian Gulf in Modern Times: People, Ports, and History*, New York, pp. 231–66.

Oppenheim, M. von (1939), *Die Beduinen unter Mitarbeitung von Erich Bränlich und Werner Caskel*, Leipzig.

Pattison, M. (1892), *Isaac Casaubon*, Oxford.

Pennington, B. K. (2005), *Was Hinduism Invented? Britons, Indians, and Colonial Construction of Religion*, New York.

Pourjavady, R. (2017), 'George Strachan', in Thomas, D. and Chesworth, J. (eds), *Christian–Muslim Relations: A Bibliographical History*, vol. 10, Leiden, pp. 565–8.

Proot, G. (2018), 'Prices in Robert Estienne's Booksellers' Catalogues (Paris 1541–1552): A Statistical Analysis', in Granata, G. and Nuova, A. (eds), *Selling and Collecting: Printed Book Sale Catalogues and Private Libraries in Early Modern Europe*, Macerata, pp. 192–221.

Rao, R. P. (1963), *Portuguese Rule in Goa: 1510–1961*, New Delhi.

Rassem, M. and Stagl, J. (eds) (1994), *Geschichte der Staatsbeschreibung: Ausgewählte 1456–1813*, Berlin

Richard, F. (1990), 'Carmelites in Persia', *Encyclopaedia Iranica*, vol. VI, 7, New York.

Robins, N. (2012), *The Corporation that Changed the World*, London.

Roncaglia, M. (1956), *Materiali per la storia del Convento di T.S. (Terra Sancta) in Aleppo*, Cairo.

Ross, E. D. (ed.) (1933), *Sir Anthony Sherley and his Persian Adventure*, London.

Rossi, E. (1953), 'Pietro Della Valle orientalista Romano (1585–1652), *Oriente Moderno* 33: 49–64.

Rothman, E. N. (2009), 'Interpreting Dragomans: Boundaries and Crossings in the Early Modern Mediterranean', *Comparative Studies in Society and History* 51(4): 771–800.

Rothman, E. N. (2012), *Brokering Empire: Trans-imperial Subjects between Venice and Istanbul*, Ithaca, NY.

Ryu, C. (2009), 'The Politics of Identity: William Adams, John Saris and the English East India Company's Failure in Japan', in Singh, J. (ed.), *A Companion to the Global Renaissance*, Oxford, pp. 178–89.

Said, E. (1978), *Orientalism*, New York.

Sale, G. (1734), *The Koran, Commonly Known as the Alcoran of Mohammed*, London.

Sanderson, M. (2004), 'William Douglas (c. 1540–1606)', *Oxford Dictionary of National Biography*, Oxford.

Summers, M. (trans.) (1971), *The Malleus Maleficarum of Heirich Kramer and James Sprenger*, New York.

Sutton, J. (2000), *Lords of the East*, London.

Swartz, F. (2017), *Writing in the Margins of Empires. The Ḥusaynābādī Family of Scholiasts in the Ottoman-Safavid Borderlands*, Vienna.

Sykes, P. M. (1915), *A History of Persia*, London.

Versteegh, K. (1997), *Landmarks in Linguistic Thought*, vol. 3, London.

Vitkus, D. (2008), '"The Common Market of all the World": English Theater, the Global System, and the Ottoman Empire in the Early Modern Period', in Deng, S. and Sebek, B. (eds), *Global Traffic: Discourses and Practices of Trade in English Literature and Culture from 1550 to 1700*, New York, pp. 19–37.

Wessels, C. (1924), *Early Jesuit Travellers in Central Asia 1603–1721*, The Hague.

Winstone, H. V. F. (1984), 'George Strachan, 17th Century Orientalist: Plea for a Biographical Study', *Proceedings of the Seminar for Arabian Studies*, vol. 14, pp. 103–9.

Winters, R., Hume, J and Leenstra, M. (2017), 'A Famine in Surat in 1631 and Dodos in Mauritius: A Long Lost Manuscript Rediscovered', *Archives of Natural History* 44(1): 134–50.

Wolff, A. (2003), *How Many Miles to Babylon? Travels and Adventures to Egypt and Beyond, from 1300 to 1640*, Liverpool.

Yule, Sir H. (1888), 'Concerning Some Little Known Travellers in the East', *Asiatic Quarterly Review* 5: 312–35.

Zarinebaf, F. (2011), 'Rebels and Renegades on Ottoman-Iranian Borderlands: Porous Frontiers and Hybrid Identities', in Amanat, A. and Vejdani, F. (eds), *Iranian Identity and Modern Political Culture*, London, pp. 81–99.

Zwartjes, O. (2012), 'The Historiography of Missionary Linguistics', *Historiographia Linguistica* 39(2–3): 185–242.

Web Addresses

Edinburgh and Toronto Universities Study (2018), available at <http://www.ed.ac.uk/news/2018/parasite-study-could-aid-efforts-to-treat-malaria> (last accessed 28 December 2018).

El País (2013), available at <https://elpais.com/cultura/2013/04/29/actualidad/1367255987_780232.html> (last accessed 3 March 2020).

Županov, I., 'The Historiography of the Jesuits Missions in India (1500–1800), *Jesuit Historiography Online*, available at <https://referenceworks.brillonline.com/entries/jesuit-historiography-online/the-historiography-of-the-jesuit-missions-in-india-1500-1800-COM_192579> (last accessed 11 January 2019).

Broadcast

Mansel, P., *The Forum*, BBC World Service, broadcast on 27 May 2016.

Index

Abbas I, Shah (of Persia), 61, 80–2, 103–10, 124–5, 128–42, 154
abbasi (silver coin), 109, 110
Abercrombie, Robert, 10–11, 22, 25–7, 29–31, 122, 143
Abī Bekr, Abdaljabbār ibn, 92
Abyssinia, 147
Afghanistan, 80, 104, 106, 125
Africanus, Leo, 4, 154
  Description of Africa, 93, 154
aga, 57, 68, 99–100, 143
Agra, India, 98, 133, 140–7, 150–1
Ahmed I, Sultan (Ottoman Empire), 57
Aix-en-Provence, France, 39, 43, 153
Akbar, Great Moghul, 143, 146
album amicorum (book of friends), 7, 19–21
  Ancina, Giovenale, 33
  Coignet, Jerome, 34
  doctors, 60, 63
  education, 12–15
  no contributions from Jesuits, 28
  Scot, Alexander, 23
  Seton, Alexander, 30
  shipwreck, 30–1
  Stichel, Patrick, 31
  travels in Italy, 25
al-Din II, Fakhr, 57
Aleppo, Syria, 53–65
  books kept in, 89
  buying books in, 83
  Della Valle, 45
  Dempster, 47–8
  Franciscan convent, 77, 129
  Levant Company, 101–2
  medical books from, 92
  missionaries in, 73
  Strachan in, 68
  'Turkey merchants', 99–100
  United Provinces of the Netherlands consulate in, 58
  Venetian postal system, 46, 58–9, 68, 101–2
  Venice consulate in, 58
Alexandria, Egypt, 46
Al-Firuzbadi, *Al-Qamus al-Muhit*, 132
Alfonso X, (of Castile), 2
al-Ḥasan, Ibrāhīm ibn, 87, 96, 98–9
al-Ḥasan at-Ṭabarsī, Raḍiyyaddīn al-Faḍl ibn, 96, 154
al-Jurjānī, Alī ibn Muḥammad, 141
Allahverdi Khan, 105, 124, 125
al-Maybudī, Mīr Ḥusayn, *Jām-i gītī-numā*, 141–2
al-Nisa Begum, Khayr, 104
Āna, Syria, 62, 66–8, 71
Anatolia, 106
Anazzah Bedouins, 66–77, 86–7
Ancina, Giovenale, 33
Andrade, Fr Antonio de, 149–50
  Relacam da Missam de Tibet, 150
an-Nawājī, *Ḥalbet al-kumyat*, 92
Anne, Queen (James VI's wife), 22, 122
apostolic nuncio, 144
Aquaviva, General Claudio, 23, 24–5, 29, 31–3, 143
Aquaviva, Rodolfo, 143, 146
Arabic language
  Bedouin tribes, 76
  books to teach with, 91
  dialects, 71
  Marracci, Ludovico, 96–7
  in Persia, 78
  Raimondi's Arabic Gospels, 72–4
  Strachan chance to learn, 55–6, 82–3
  Strachan's legacy, 153–4
  trade in Persia, 117
Arabic literature, 81, 92
Arabic poetry, 93–4, 154
Armenia, 80
Armenians, 108
  Orthodox priests, 131
  silk trade, 110
Ashrafiya madrassa, Damascus, 88–9
Asia, 130
Asia, Central, Jesuit exploration in, 147–9

Asia Minor, 52
Aṣ-Ṣaḥīfa al-kāmila, 87, 96
Augsburg humanists, 35
Augustinians, 129–30
Aurangzeb, 147
Azevedo, Don Jerónimo de, 106–7
Azevedo, Fr Francis de, 150

Badrinath, 149
Baffin, Captain William, 127
Baghdad
  buying books in, 83, 89
  capture of, 56, 79–80, 80, 128, 139
  Della Valle, Pietro, 46
  East India Company, 102
  libraries, 154
  Nellson, 98–100, 106, 129
  Strachan books found in, 141
  Strachan escape to, 87
  studies in, 95
Bahrain, 106
bailo (Venetian ambassador), 3, 54
Bandar Abbas *see* Gombroon, Persia
Barbary Company, 100
Barberini, Maffeo, 34–8
Barclay, John, 14, 35, 39
Barclay, William, 35
Barker, Robert, 113–15, 117, 121
Barker, Thomas, 102–3, 112, 115, 119, 154
Beaton, James, 10–11, 11
Bedouin emirs, 61–5
Bedouin tribes, 58, 61–3, 66–77, 154
Beg, Huseyn Ali, 125
Beg, Uruch, 125
Bell, William, 118, 124, 139–40
Bengal, 146–147, 151
Bengal Army, 72
Benthall, Captain John, 137
berat (certificate), 54
Bernardo, Lorenzo, 54
Blackwood, Robert, 28
bookbindings, 90
Borghese, Camilo, 24, 34, 36
Borromeo, Cardinal Federico, 24
Bosworth, Professor, 1
Brahmins, 145
Braunsberg, 29
Brechin Cathedral, 28
British East India Company, 72
Brudel, Fernand, 40
Buckingham, Duke of, 127–8
Buddhism, 149
Burnet, Gilbert, 20
Byzantine Empire, 41

Cabbala, 4
call to prayer, 84
Calvinism, 17, 20, 21, 25–7, 76
  Calvinist Confession of Faith, 16, 24
Cape of Good Hope, 41
caravans, 61–2, 66, 69–71
Cardowe, Mr, 119–21, 122, 123, 139
Carmelite friars
  books entrusted to, 93, 96, 134
  Strachan's Catholicism, 146
  Strachan in contact with, 140
  Strachan's will held by, 153
  taught Arabic by Strachan, 74
  see also Discalced Carmelite Friars, Isfahan
Carmelite Order, Rome, 153
Carpentras, Venaissin, Rhone Valley, 21, 25
Casaubon, Isaac, 36
Casaubon, John, 36–7
Cathay, 148–9
Catherine de Medici (of France), 55
Catholic Church, Eastern languages and culture, 5
Catholic religious orders, 59–60
Catholicism
  Aleppo, 59–60
  Casaubon, John, 36
  Charles, Prince (of England), 122
  colleges, 29–30
  de Sancy, 53
  Della Valle, 133
  East India Company, 103, 121, 126
  Henri IV (of France), 15
  James VI (of Scotland), 21–3
  James VI/I (of Scotland/England), 37–8
  Jerusalem, 49–50
  Parent, Signor Alviso, 119
  Portuguese colonies, 148
  religious orders in Persia, 129–31
  in Scotland, 8–11, 15–17, 28
  Strachan and the Bedouins, 72–4, 76, 85
  Strachan and the Portuguese, 146
  Strachan family, 27
  universities and, 20–1
  Virgin Mary, veneration of, 13–14
Cattaro, Venetian Republic, 46–7
Caucasians, 104–5

Caucasus, 56, 106, 108
Cecil, William, Lord Burghley, 9–10
Çelebi, Katib, 82
Çelebi, Mohammed, 66–77, 85–6
Central Asia, Jesuit exploration in, 147–9
Chaldean language, 50–1
Chalmers, Thomas, 14
Chambers, David, 36, 37
Charles, Prince (of England and Scotland), 122, 127–8
Charles I (of England and Scotland), 139–40
Charles IX (of France), 10
Charles of Lorraine, 4th Duke of Guise, 38–9, 64
Charles V, Emperor, 50
Châtel, Jean, 15
Cheuart, Pietro, 123
China, 148–9, 152
  silk, 41–2
Chinese traders, 149
Chisholm, Bishop William, 21–3, 26, 37
  Examen Confessionis Fidei Calvinianae quam Scotis Subscribendum proponent, 21
Christian traders, 2–4
Church of the Apostle Thomas, 144, 148
Church of the Holy Sepulchre, Jerusalem, 48–50
Churches of the Eastern Rites, 51, 55, 148
cinchona, 65, 116, 119
Claremont College, Paris, 11–12, 15
Clavius, Christoph, 2
Clement VIII, Pope, 21–2, 23, 32
Coignet, Jerôme, 34
Collège du Mans, 34, 35, 36
College of Cardinals, 21–2
College of La Rochelle, 20
College of St Pancratius of the Discalced Carmelites, 96–7
Combru, Gombroon, 72
concertatio, 83
Constantinople, 44–8, 52–5, 99, 108, 131
  training school, 3
Convento di Terra Santa, 59, 129
Coresma, Fr Nuño, 150–3
Coryate, Thomas, 69
Cotton, Sir Dodmore, 125
Council of Trent, 143–4
Coutinho, Don João, 106–7
Crichton, William, 10, 11
Curia, 22, 25
Cyprus, 42

Dalrymple, William, 50
Dāmād, Mīr, 141, 150, 154
Dara Shukoh, 147
Datary, Cardinal, 37
de Silva y Figueroa, Don Garcías, 103–4, 106–7, 131
de Vega, Lope, *La Corona Tragica*, 146
De Young, Gregg, 141
Deays, 55
Delhi, India, 72
Della Valle, Manni, 126–7, 132, 135–6
Della Valle, Pietro
  Abbas I, 107
  Aleppo, 60
  all Muslims called Turks, 49
  alla moresco (disguise), 70
  belief Strachan was dead, 140–1
  Carmelite friars, 124
  Constantinople, 56
  de Sancy, 53
  de Silva, 104
  disease, 63
  Egypt, 45
  East India Company, 139
  Feyyād, Emir, 64–5
  friendship with Strachan, 131–7
  Gombroon, 126
  Jerusalem, 49
  journey of, 44–6
  meeting with Strachan in Persia, 153
  Quli Khan, 135–7
  sea journey, 47
  as source of information, 7
  Strachan as Bedouin, 155
  Strachan's desert life, 66–77
  Strachan's discourses with teachers, 82–5
  Strachan's escape from Feyyād, 86
  Strachan's Trivium, 12
  visiting Classical sites, 51–2
  wife's fatal illness, 126–7, 135–6
Dellavida, Giorgio Levi
  biography of Strachan, 1, 6
  East India Company, 124
  Strachan and medical books, 91–2
  Strachan manuscripts, 141, 153
  Strachan's proficiency in Arabic, 94–5
  Wiest, 30

Dempster, Thomas, of Muiresk
biography of Strachan, 43, 137
Catholicism, 20
Collège du Mans, 34
Historia Ecclesiastica Gentis Scotorum, 7
James VI/I, 38–9
letters to, 59, 68
in need of sponsorship, 19
praise for 'Lacrymae', 35
Strachan's chance to learn Eastern languages, 76
de'Pizzicoli, Cyriaco, 44
Dervish Celali tribesmen, 80
Desideri, Ippolito, 152
Discalced Carmelite Friars, Isfahan, 89, 103, 119, 122, 123, 130–1, 137; *see also* Carmelite friars
doctors, 62–5
Douai, Spanish Netherlands, 29
Douglas, Sarah *see* Strachan, Sarah (sister-in-law)
Douglas, William, 10th Earl of Angus, 15–16
dragomans, 54–5, 139
Druze emirs, 61–2
Du Perron, Cardinal Jacques Dary, 37–8, 59
du Prat, Bishop William, 11
Dunottar Castle, 8
Dupuy, Christophe, 58–9, 98
Duras, Fr George, 25
Dutch East India Company (VOC), 58, 110, 138–9
Dutch merchants, 151

East India Company, 98–128
archive, 7
Barker about Strachan, 63, 155
Della Valle, 133, 135
Massacre at Amboyna, 138–9
Strachan and the Portuguese, 146
Strachan's income from, 90–1
Strachan's marriage, 76
Strachan teaching Arabic, 74
Strachan's letter to Smyth, 153
Trading Factory, Isfahan, 102–4
Eastern languages
Aleppo, 60
Bedouin tribes, 76–7
de Sancy, 53, 55–6
Jerusalem, 51–2
lack of proficiency in, 2–4
Strachan's wish to learn, 43
teaching of, 154
trade, 59
Eastern literature, 58–9, 82, 140
Eastern Rites Christians, 52
Edinburgh, 26
Egidio da Viterbo, Cardinal, 4
Elizabeth I (of England), 11, 100
Elphingston, Alexander, 18
English trading companies, 100–2; *see also* English East India Company; Levant Company
Erpenius, Thomas, 154
Oration on the Value of the Arabic Language, 93
Rudimenta linguae Arabicae, 93
Ethiopia, 147
Euclid, *Elements*, 90–1, 141
Eugenius IV, Pope, 44
Eyre, Sir John, 99

Fakhr al-Din II, 62
famine, 151–3
Farsi language, 55–6, 78, 117, 130
fasting, 84
Ferdinand of Tuscany, Duke, 57
Feyyād Abū Rīsha, Emir, 61–77
Della Valle, 131, 133
Ḥusaynābādī scholiasts, 81–3, 155
and sexual continence, 64–5
Strachan and the Portuguese, 146
Strachan books collected whilst in service to, 95
Strachan's escape from, 85–6, 98–9
Strachan's finances, 89, 91, 116
Strachan's medical books, 92
Strachan's New Testament, 151
Ficino, Marsillio, *De Vita*, 64
Fleet prison, 113–14
Fleming, James, 20
Florence, Italy, 57
Fracastoro, Girolamo
De Contagione et Conagiosis Morbis, 65
'Syphilis sive morbus gallicus', 65
Franciscan Order
Aleppo, 59–60, 68, 129
Fakhr al-Din II, 57
French, 48–50
Jerusalem, 52, 143–4
medical treatment, 63
Raimondi's Arabic Gospels, 73
Franks, Western Christians, 49
Fraser, Mr, 37

French Compagnie des Indes Orientales, 110
Fück, Johann, 1

Gabriel, Juan, 4, 154
Galilee, 57
Geneva, Switzerland, 30, 31
Genoa, Italy, 2
Georgian Orthodox priests, 129–30, 131
Ghetaldi, Marino, 23
ghulams (servant of the shah), 104–6, 108, 124, 125
Gibbon, Edward, 97
Gifford, Robert, 103, 126, 132
Giorido, Abdullah, 132
Giovanni, Fr, 132
Glasgow, archbishop of, 10–11
'Global Village', 40
Goa, India
  Andrade, 150
  Azevedo, 106
  Coresma, 152
  de Silva, 103–4
  Della Valle and Strachan, 135
  Inquisition, 144–6, 150
  Jesuits, 143
  missionaries, 129
  Portuguese, 128
  Strachan, 153
  Xavier, 144
Goes, Bento de, 148–9, 152
Golden Horn, 52, 54
Golius, Jacobus, 93–4, 154
Gombroon, Persia, 126, 128, 136–7, 138, 140
Gordon, George, 6th Earl of Huntly, 15
Grand Tour of the East, 45
Great Moghul, 98, 102, 129, 133, 140
Greek Orthodox Church, 50
Gregory XIII, Pope, 10
Grimani, Marino, 32
Guadagnoli, Filipo, *Breves arabicae linguae institutiones*, 94
guaiacum, 65, 116
Guéret, Fr, 15
Guignard, Fr Jean, 15
Guise, 4th Duke of, Charles of Lorraine, 38–9, 64
Gunpowder Plot, 36, 122
Gustavus IV Adolphus (of Sweden), 10

Ḥāḍir, 69, 71
Ḥadīths, 95
Hajj pilgrimage to Mecca, 45
Ḥalbet al-kumyat, 95, 120–1
Hamilton, John, 20
Hamilton, Sir Thomas, 16
Harlay, Achille de, Baron de Sancy, 53–6, 131
Harper, James G., 40
Hay, Edmund, 11, 15
Hay, Francis, 9th Earl of Erroll, 15
Haydar, 82
Haydarani, Ahmad, 81–2
Haywkyns, John, 114
Hebrew language, 50–1
Henri III (of France), 10
Henri IV (of France), 10, 15, 36, 37–8, 53, 59
Herat, Aghanistan, 106
Hinduism, 143–5, 149
Historiographer Royal, 38
History of Oriental Studies, The, 1
Hoeschel, David, 35
Holy City, 48–51
Holy Land, 47–8
Hormuz, Persia
  capture of, 124–8, 136, 138
  Jesuit missionaries, 129
  Strachan and attack on, 133, 135, 154
Hubert, Etienne, 55, 154
Huguenots, 21
Ḥusaynābādī scholiasts, 78–87, 105, 154

IHS symbol, 34
Il Gran Delfino (ship), 47
I'lām al-warā bi-a'lām al-hudá, 96, 97, 154
ilitizam, 56, 61–2
Ilkhan, Ghazan, 79
India
  Aleppo, 59
  caste system, 145
  Della Valle, 135, 137
  East India Company, 111
  Jesuits arrival in, 143–4
  Ochterlony, 72
  spices, silks, and cottons, 41–2
  Strachan's intention to travel to, 129
Indonesia, 111
Inquisition
  Goa, 144–6
  Spain, 3

Iran
  Ḥusaynābādī scholiasts, 81
  silk trade, 108–9
  state expenditure, 107–9
  Strachan in, 90–1, 150, 155
  'the third force,' 105, 131
  trade, 137–40
  see also Persia; Safavids
Iraq, 154
'Iron Curtain' model, 40
Isfahan, Persia
  Abbas I, 104–9
  Augustinians, 129–30
  Carmelite friars, 74, 134, 146, 153
  Della Valle, 46, 137
  East India Company, 63, 102, 155
  silk trade, 61
  Strachan buying books in, 90
  Strachan in, 119, 150
Islam
  and Christianity, 40–2
  confessional identities in Iran, 78–80
  conversion, 151
  doctors, 62–3
  'false religion', 3–4, 74, 96, 154
  'Five Pillars of Islam', 84
  Jerusalem, 48–52
  polygamy, 4
  as religion of lust, 3–4
  seclusion of women, 4
  'Turks', 49
Islamic Arabic texts, 2–4
Islamic studies, books on, 92
Ismail I, Shah, 79–80
Ismail II, Shah, 105, 110
Italy
  language schools, 5
  silk, 41

Jahan, Shah, Great Moghul, 147, 150–1
Jahangir, Great Moghul, 146
Jām Gētī Numā, 150
James I (of Scotland), 8, 27
James VI (of Scotland)
  Beaton, 10–11
  Chisholm, 21–2
  relations with Spain, 15–16
  Strachan and, 26–8
James VI/I (of Scotland/England)
  dragomans, 54
  East India Company, 128, 139
  and 'false religion', 74
  relations with Spain, 99
  Strachan and, 36–9
Japan, 148
Japanese traders, 138–9
Jasques, Persia, 120, 124, 127
Jefferson, Thomas, 97
Jeffris, Robert, 113–14, 119–24, 127, 133
Jerusalem, 4, 47–52, 143
  books about pilgrimages to, 44
  Jewish Quarter, 51
  pilgrims, 50
  sacked by Bedouins, 62–3
Jesuit archives, Rome (ARSI), 7
Jesuit missionaries, 22, 28, 73–4, 129
  in Central Asia, 147–54
  conversion of the elite, 73–4
  in India, 143–7
  letters, 22
  in Scotland, 9–11, 28
  Stichel, 31
  Strachan as courier for, 24–6
Jesuit Spiritual Exercises, 15
Jesuits
  Aquaviva, General Claudio, 23
  arrival in India, 143–4
  cinchona, 65
  Clavius, 2
  de Sancy, 53, 55
  education, 11–12, 73
  exploration in Central Asia, 147–9
  Henri IV (of France), 14–15
  at the Moghul court, 146–7
  in Portuguese garrisons, 129
  Ratio Studiorum, 51, 83–4, 91
  Regensburg, Bavaria, 33
  in Scotland, 15–16
  Scottish, 11–12
  Strachan, 29–31
  Strachan and Barberini, 37
  Strachan distancing himself from, 20–1
  Strachan family, 15–18
  Venetian Republic, 32
Jews, 51, 59
Jibrin, Syria, 70
Joao III (of Portugal), 144
Julfna, Armenia, 108
Justinian I, Emperor, 41

kafirs, 49, 62, 77, 82, 85, 95, 100
Kandahar, Afghanistan, 128
Keith, Isobel, 8

Khan al-Burghul, Aleppo, 58
Khan al-Gumru, Aleppo, 58
Khan al-Shouneh, Aleppo, 57–8
khans (fortified inns), 57–8
King's College, Aberdeen, 9
Kirmān, Persia, 108
Kishm, Persia, 103, 107, 127–8, 129, 138
Konkani language, 145
Kufi script, 78
Kuhestak, Persia, 125, 127
Kurdistan, 80
  schools, 81

Ladakh, Tibet, 149–50
Ladino language, 51
Lahore, India, 146
Lamas from the West, 149–50
Lar, Persia, 137
Lars, Persia, 108
Lebanon, 56–7
letters of credit, 110
Levant, 44, 47–8, 51
Levant Company, 54, 58, 99, 100–1, 110–11, 120
lingua franca, 2–3, 59
Lipsius, Justus, 20
Lithgow, William, 39, 43, 44, 52, 153
Louis XIII (of France), 58
Loyola, Ignatius, 143–4
Lubnān, Jabal, 56–7
Lucknow, India, 72

Ma'ānī al-āthār, 95
Ma'anid family, 56
MacQuhirrie, Alexander, 22–3
Majlis Shūrāh Library, Tehran, 141
Makcall, Prior Adam, 33
malaria, 65
Malleus Maleficarum, 'The Hammer of Evil-doers,' 74
Malpichi, Fr Stanislaus, 147
Mamluk Egypt, 4, 61, 78, 80
  market for silk and woollen textiles, 41
Manuel I Komnenos, 147
manuscripts, cost of, 89–91
Maria Anna, Infanta (of Spain), 122, 127–8
Marie de Medici (of France), 53
Marischal, Earl, 8
Maronite Christians, 57
Marques, Brother Manoel, 149, 151
Marracci, Ludovico, 96–7, 154
Mary, Queen of Scots, 8, 10, 11, 146
Massacre at Amboyna, 138–9
Mawālīs (converts to Islam), 78
McGhie, Johne, 16
McRoberts, Fr David, 1
Mediterranean Sea, 41, 52
Meir, Albrecht, *Methodus desribendi regions, urbes et arces,* 45
Melanchthon, Philipp, 12–13
Melluha, Syria, 70
Membré, Michel, 55
mensam ambulatorium (the walker's table), 43
merchants, 59–61, 69–71, 98, 110, 125–6, 140
  Dutch, 151
  Kashmir, 149
  'Turkey merchants', 99–100, 101–2, 106
Meydān-e Mir, Isfahan, 130
Middle East, Lithgow, William, 39
Minab, Persia, 126–7, 135
missionaries, 5, 129–31
Moghul court, 72, 111, 143, 145–6
  Jesuits at the, 146–7
Mohammad I, Shah, 104, 109
Mongols, 78–9, 148
Monnox, Edward, 102–3, 113–14, 117–24, 126–7, 139, 154
Montauban, France, 21
Mount Lebanon, 44, 48, 56, 57
Mu'allaqāt, 102
Mudlij, 62
Murshid Qoli Khan, 104
Muscat, Oman, 128, 136
Muslims *see* Islam

Naghsh-i Jahan, 108
Naples, Italy, 57
Nellson, William, 98–100, 106, 116, 129
Neo-Stoicism school, 20
Nestorians, 148
'New Christians', 144
New Julfna, Armenia, 108, 131
North Africa, 55, 100, 154
Northern Collge, Braunsberg in Livonia, 9–10
Nusseibeh family, 50

Ochterlony, General David, 72
Oghuz Turks, 79–80
opium, 116
Order of the Caracciolini, 5, 154

Order of the Clerks Regular of the Mother of God of Lucca, 96
Oriental Press, Rome, 73
Orientalism, 40
Orthodox Church, 52
Osman, Sari, 81–2
Ottoman Empire
  Abbas I, 81, 103–9, 125, 138
  Aleppo, 56–60
  Augustinians, 130
  Bedouin tribes, 61–2
  caravans, 70
  communication, 68
  Constantinople, 53–4
  Della Valle, 45–8, 49
  and Eastern Christians, 143
  East India Company, 101–2
  literature, 83, 89
  Nellson, 99–100
  Qizilbash (Red Hats), 80
  trade, 2–3, 42, 61, 110
  Turkish language, 55–6
  Venetian Republic and, 4
Ottoman sultans, 56
  Persia, 56
Ottoman Turks, 61

Pahlavi language, 78
Palmyra, Syria, 62
Paolini, Mgr Bernardino, 24
Papal States, 32
papermaking, 88
Paṛaiyār (pariahs), 145
Paravars, 145
Parent, Signor Alviso, 119
Paris, France, 25, 31, 33–6
Patten, Eduard, 114, 120, 121
Paul III, Pope, 143–4
  Licit ab initio, 144
Paul V, Pope, 5, 24, 34, 36–7
Pavia, Italy, 24, 25
Penal Laws (Scottish), 10, 16, 17
Pera, Constantinople, 53–4
peregrinatio academica, 13
Persia
  Catholic religious orders, 129–31
  Ottoman sultans, 56
  trade, 107–9
  see also Iran
Persian language, 72, 147
Persian literature, 81, 92
Persian State archives, 140–1
Philip II (of Spain), 15
Philip III (of Spain), 103–4, 106–7
Philipp II/III (of Portugal/Spain), 130
Phillipp III (of Spain), 16
Photius, *Bibliotheca*, 35
Piedmont, Italy, 33
pilgrimages, 40, 44, 47, 149
Pindar, Paul, 99
Pius IV, Pope, 145
Pius V, Pope, 130
poetry, Arabic, 92–4, 154
Pont-a-Mousson in Lorraine, college, 35
Pontifical Scots College, Rome, 22, 29–30, 31, 43
Port Hooghy, Ganges delta, 151
Portuguese
  Abbas I, 103–4, 106–7, 134
  Della Valle, 136
  Hormuz, 124–8
  Inquisition in Goa, 144–6
  Jahan, Shah, 147, 151
  missionaries, 130
  Prester John, 147–8
  Strachan's New Testament, 152
  trade, 110–11
  trade routes, 61
Pourjavady, Reza, 141
prædominium (guardianship), 50
Presbyterians, 26, 27, 28
Prester John, 147–9
Privy Council (Scottish), 18, 26–7
Propaganda Fide 1622, 130
'propagation de la foy, la' ('spreading the faith'), 58
Prospero of the Holy Spirit, Fr, 134, 136
Protestant University of Béarn, Lescar in Navarre, 14
public libraries, 88–9

Qazvin, Iran, 107–8
Qizilbash (Red Hats), 80–1, 104–6, 109
Quadrivium, 12, 14, 19
Queiro, Fernao de, 153–4
Quince, Thomas, 123–4
Quli Khan, Imam, 124–8, 136, 140, 154
Qur'an
  Arabic language, 71, 78
  Arabic poetry, 93
  kafirs, 85
  Latin translation, 154
  Marracci, 97
  mistranslated, 3–4

slavery, 41
Viterbo, 4

Raimondi, John Baptist, 73
Arabic Gospels, 72–4
Rait, Magdalene (Strachan's sister), 30
Rait family, 30–1
Ratio Studiorum, 12, 51, 83
Regensburg, Bavaria, 33
Reid, Thomas, 20
religious tolerance, 78–9, 145
Republic of Letters, 1–2, 35
Ricci, Matteo, 148–9
Rich, Claude, 141
Risāla-I Mu'īniyya, 91
Robbins, William, 111–12
Robert of Ketton, 3
Robins, William, 102–3
Roe, Sir Thomas, 111
Rollok, Walter, 30
Rosinus, Johannes, *Antiquitatum Romanorum Corpus Absolutissimum*, 38
Rosweyde, Herbert, 36
Rynns, William, 121

Sacrae Linguae, 51
saettia, 52
Safavayga, 79
Safavids
Augustinians, 130
de Silva, 104
Feyyād, Emir, 61–2
finance, 109
Qizilbash (Red Hats), 80
relations with Spain, 106–7
and religious tolerance, 79
scholars, 77
Sherley brothers, 125
Sa'ib of Tabriz, 109
Said, Edward, 40
Sale, George, 97
Sandys, George, 48, 49, 50, 51
Sant Jazques, Estefano de, 123
Santa Maria dell'Umilità, 32
Sarab, Persia, 106
Saville, Henry, 99–100, 101–2
Schiapano, Doctor Mario, 45, 153
Schott, Andreas, 35
Schottenkloster, Regensburg, 43
Scot, Alexander, 23
Scots College, Paris, 36, 43
Scots College, Pont-a-Mousson in Lorraine, 10–11, 11
Scots College, Rome, 23–4
Scots Colleges, Louvain, 20
Scottish Benedictine monastery of St James, 33
Sea of Marmara, 52
Seget, Thomas, 20
Selim I, Sultan (Ottoman), 4, 80
Sephardic Jews, 51
Seton, Alexander, 122
Seton, Lady, 29–30
Seton, Patrick, 29–31, 35
Sevenduk Khan, 136, 140
shahi (coin), 109
Sherary, 114
Sherley, Anthony, 125
Sherley, Robert, 125
Sherley, Thomas, 125
Shi'a Islam
Abbas I, 81, 131, 138
and religious tolerance, 79
slavery, 41
Strachan learning about, 96
'the third force', 105
Shiraz, Iran, 118–19, 124, 126, 129–30, 136
Sibawayh, 78
Sidon, Lebanon, 57
signoria (governing body), 3, 32, 46, 54–5
Silk Road, 58, 60–1, 148–9, 149, 152
silk trade, 41–2, 110–12, 128
Moriscos, 41
slavery, 40–1
Smyth, Sir Thomas, 100, 114–17, 129, 142, 153
Souk al-Madina, Aleppo, 58
Souk al-Zirb, Aleppo, 58
souks (large covered marketplaces), 58
Spain
Azevedo, 106–7
de Silva, 103–4
Inquisition, 3
missionaries, 130
relations with England, 99
silk trade, 41
trade with Asia, 110
Spanish Blanks, 15–16
Spice Islands, 111, 120
Sringar mission, Garhwal, 143–56
St Paul's Collge, Goa, 144, 146, 153–4

St Saviour convent, Jerusalem, 48–9, 143
Stafford, Sir Edward, 11
Stammbuch, 12–13
St-Elise, Fr Jean-Thaddée de (Fr Giovanni), 130
Stichel, Patrick, 31
Stinson, Andrew, 25, 27
Strachan, Alexander (nephew), 9, 16, 17–18, 24–5, 139–40
Strachan, George
  accused of poisoning, 121–4
  background, 5–7
  as a Bedouin, 71–2, 85–7
  books, access to, 88–9
  books, collection and translation of, 5–6, 76–7, 88–91
  books, purchase of, 69, 83
  catalogue of books, 89
  change of name to Mohammed Çelebi, 75
  circumcision, 86
  contribution to the European study of Islam, 95–7
  conversion to Islam? 75–7, 83–7
  death, 152–3
  in the desert, 66–77
  disillusionment, 36–8
  dismissal, 121–4
  as doctor, 63–5, 69, 111–12, 116–17
  'Doctor Mohammed', 75
  educated by Jesuits, 11–12, 83–4
  education, higher, 9–11
  escape from Feyyād, 84–7
  exiled, 27–8
  family debts, 18
  family difficulties, 15–18
  'Faustissimam Sancissimi D(omini) N(ostri) Pauli V Pont. Max. inaugurationen, orbi christiano gratulatur Georgius Strachanus', 34
  final journey, 150–3
  financial circumstances, 19–20
  friendship with Della Valle, 131–7
  as godfather, 132
  as independent trader? 137–40
  intellectual debate, 83–4
  as interpreter, 112, 124–6
  as Jesuit courier, 24–5
  Jesuit leanings, 17, 29–31
  'Lacrymae', 30–1, 35
  legacy, 153–6
  letter from home, 24–6
  library, 88–97, 134–5, 153, 154–6
  malaria, 63, 118–19, 137
  marriage, 75–6, 86, 98
  'Master of the Rod', 66–7
  mathematics, 91
  medical books, 91–2
  New Testament, 151–2
  peregrinations, 19–21
  personal seal, 13
  poetry, 6–7, 8, 24
  proficiency in Arabic, 94–5
  of royal descent, 8–11
  search for a patron, 33–6
  as Shah Abbas's divan, 140–2
  shipwreck, 30
  teaching Arabic, 134
  teaching languages, 120–1
  translations, 34–5
  travelling to the Ottoman Empire, 46–8
  on trial for treason, 26–8
  visiting Classical sites, 51–2
  will, 135, 153
Strachan, John (brother), 16–17, 24, 26, 27
Strachan, Robert (brother), 8–9, 17
Strachan, Roger, of Glethknow, 18
Strachan, Sarah (sister-in-law), 8, 16, 17, 24–5
Strachan, Sir Alexander (father), 8, 16
Strachan family, conversion to Calvinism, 24–7
Straits of Hormuz, Persia, 61, 107
Sublime Porte (Âsistâne-yi Sa'âdet), 4, 55, 80, 138
Sufism, 79
Suleiman the Magnificent, Sultan (Ottoman), 48–50, 51, 56, 80
Sunni Islam
  Abbas I, 105
  Akbar, Great Moghul, 146
  kafirs, 62
  and religious tolerance, 79–81
  scholars, 77
  Selim I, 4
  slavery, 41
  Strachan learning about, 96
Surat, India
  Della Valle, 136
  East India Company, 111, 120, 139
  famine, 151

Jeffris, 123–4, 127
Portuguese, 124
Sylvester II, Pope, 2
syphilis, 65
Syria, 78, 154
Syriac Christians, 59, 129–30
Syrian Desert, Great, 61–2, 65, 66–7, 131
schools, 81

Tabriz, Iran, 80, 106, 107–8
Tahmasp I, Shah, 80, 108, 109
taqiyya, 138
taxation, 70–1, 78–9, 107–9, 126, 138, 143
Tempesta, Antonio, 73
Temple Mount, Jerusalem, 50
Tengnagel, Sebastian, 137
Third Spalding Club, 6
Thirty Years' War, 139
Thornton Castle, Scotland, 8, 27
Thornton coat of arms, 13
Tibet, 149, 150, 151–2
Ticino, Italy, 24, 25
Tiepolo, Antonio, 3
Timurid rulers, 79
trade
arms and munitions, 109
drugs, 116–17
English woollen cloth, 111
Flanders woollen textiles, 41
Flemish silk, 41
between Italy and Mamluk Egypt, 2
between Italy and Ottoman Empire, 2
luxury commodities, 41–2
opium, 116
Persia, 107–9
private trading, 113–14, 119–20
routes, 57–8
Russians, 110
by sea, 110
sericulture, 41
silk, 41, 41–2, 57, 110–12, 128
silver, 110–11
spices, 57, 60–1, 111
Venetian Republic, 4
travel
accounts, 44
sea journeys, 52
as self-improvement, 42–3
Treaty of Constantinople 1590, 106, 138
Treaty of Nasuh Pasha 1612, 106, 109, 138
Treaty of Tordesilhas, 130
Treaty of Zaragoza, 130
Tri Tashi Drakpa (king of Guge), 149
Trivium, 12, 14, 91
Troy, ruins of, 39
Tsaparang, Guge, 149–52, 154
Strachan, 150
Turkey, Della Valle, Pietro, 45
Turkey Company, 100
'Turkey merchants', 99–100, 101–2, 106
Turkish language, 55–6
Turkmen tribesmen, 105
Twelver Shi'ism, 79, 79–80
Tyrie, John, 11

United Provinces of the Netherlands, 139
University of Aberdeen, 14
University of Bordeaux, 21
University of Carpentras, 23
University of Leiden, 93, 154
University of Lescar, 20–1, 63
University of Montpellier, 21, 63
University of Naples, 94
University of Paris, 19, 33–4, 38, 55, 154
University of Pennsylvania, 94
University of Rome, 94
University of Toulouse, 21
University of Turin, 94
Urban VIII, Pope, 34, 146
Urquharts of Meldrum, 17
Uzbeks, 104, 106, 107, 125

Vaison, See of, 21, 25
Valencia, Spain, silk, 41
Vatican archives (ASV), 7
Vatican library, 153
Venetian ducats, 110
Venetian postal system, 46–7
Venetian Republic
dominance of, 42–3
military leagues, 4
Most Serene Republic (La Serenissima), 4
and Ottoman Empire, 42–3
peace treaty 1573, 4
'the Turk's Courtesan', 4

Venice, Italy
  Strachan, 31–2, 46
  trade, 2–3
  Turkish merchants in, 54–5
Venice Company, 100
Vincent of Saint Francis, Fr
  Arabic poetry, 93
  letters carried by, 137
  medical books, 91
  Strachan's library, 89, 96–7, 134–5, 141, 153
Vitelleschi, Mutio, 150
Viterbo, Cardinal, 154

Wade, Sir William, 11
Whyte, Abbot James, 33
Wiest, 30
Winstone, Victor, 'George Strachan, 17th Century Orientalist: Plea for a Biographical Study', 1
Winzet, Ninian, 33

Xavier, Francis, 144, 145

Yule, Major William, 72–4, 151

Zhu Yijun, Chinese Wanli emperor, 148

EU representative:
Easy Access System Europe
Mustamäe tee 50, 10621 Tallinn, Estonia
Gpsr.requests@easproject.com

www.ingramcontent.com/pod-product-compliance
Lightning Source LLC
Chambersburg PA
CBHW070355240426
43671CB00013BA/2509
* 9 7 8 1 4 7 4 4 6 6 2 3 3 *